ACCENT ON FORM

WORLD PERSPECTIVES · *Volume Two*

Planned and Edited by RUTH NANDA ANSHEN

ACCENT ON FORM

An Anticipation of the Science of Tomorrow

By LANCELOT LAW WHYTE

New York

HARPER & BROTHERS PUBLISHERS

WORLD PERSPECTIVES: VOLUME I

Approaches to God, by Jacques Maritain

Library of Congress catalog card number: 54–6036

Contents

"*Form poses a problem which appeals to the utmost resources of our intelligence, and it affords the means which charm our sensibility and even entice us to the verge of frenzy. Form is never trivial or indifferent; it is the magic of the world.*"

A. M. DALCQ, in *Aspects of Form.*

World Perspectives

WORLD PERSPECTIVES is dedicated to the concept of man born out of a universe perceived through a fresh vision of reality. Its aim is to present short books written by the most conscious and responsible minds of today. Each volume represents the thought and belief of each author and sets forth the interrelation of the changing religious, scientific, artistic, political, economic and social influences upon man's total experience.

This Series is committed to a re-examination of all those sides of human endeavor which the specialist was taught to believe he could safely leave aside. It interprets present and past events impinging on human life in our growing World Age and envisages what man may yet attain when summoned by an unbending inner necessity to the quest of what is most exalted in him. Its purpose is to offer new vistas in terms of world and human development while refusing to betray the intimate correlation between universality and individuality, dynamics and form, freedom and destiny. Each author treats his subject from the broad perspective of the world community, not from the Judaeo-Christian, Western, or Eastern viewpoint alone.

Certain fundamental questions which have received too little consideration in the face of the spiritual, moral and political world crisis of our day, and in the light of technology which has released the creative energies of peoples, are treated in these books. Our authors deal with the increasing realization that spirit and nature are not separate and apart; that intuition and reason must regain their importance as the means of perceiving and fusing inner being with outer reality.

Knowledge, it is shown, no longer consists in a manipulation of man and nature as opposite forces, nor in the reduction of data to mere statistical order, but is a means of liberating mankind from the destructive power of fear, pointing the way toward the goal of the rehabilitation of the human will and the rebirth of faith and confidence in the human person. The works published also endeavor to reveal that the cry for patterns, systems and authorities is growing less insistent as the desire grows stronger in both East and West for the recovery of a dignity, integrity and self-realization which are the inalienable rights of man who may now guide change by means of conscious purpose in the light of rational experience.

Other vital questions explored relate to problems of international understanding as well as to problems dealing with prejudice and the resultant tensions and antagonisms. The growing perception and responsibility of our World Age point to the new reality that the individual person and the collective person supplement and integrate each other; that the thrall of totalitarianism of both right and left has been shaken in the universal desire to recapture the authority of truth and of human totality. Mankind can finally place its trust not in a proletarian authoritarianism, not in a secularized humanism, both of which have betrayed the spiritual property right of history, but in a sacramental brotherhood and in the unity of knowledge, a widening of human horizons beyond every parochialism, and a revolution in human thought comparable to the basic assumption, among the ancient Greeks, of the sovereignty of reason; corresponding to the great effulgence of the moral conscience articulated by the Hebrew prophets; analogous to the fundamental assertions of Christianity; or to the beginning of a new scientific era, the era of the science of dynamics, the experimental foundations of which were laid by Galileo in the Renaissance.

An important effort of this Series is to re-examine the contradictory meanings and applications which are given today to such terms as democracy, freedom, justice, love, peace, brotherhood and God. The purpose of such inquiries is to clear the way for the foundation of a genuine *world* history not in terms of nation or race or culture but in terms of man in relation to God, to himself, his fellow man and the universe, that reach beyond immediate self-interest. For the meaning of the World Age consists in respecting man's hopes and dreams which lead to a deeper understanding of the basic values of all peoples.

Today in the East and in the West men are discovering that they are bound together, beyond any divisiveness, by a more fundamental unity than any mere agreement in thought and doctrine. They are beginning to know that all men possess the same primordial desires and tendencies; that the domination of man over man can no longer be justified by any appeal to God or nature; and such consciousness is the fruit of the spiritual and moral revolution through which humanity is now passing.

World Perspectives is planned to gain insight into the meaning of man, who not only is determined by history but who also determines history. History is to be understood as concerned not only with the life of man on this planet but as including also such cosmic influences as interpenetrate our human world.

This generation is discovering that history does not conform to the social optimism of modern civilization and that the organization of human communities and the establishment of justice, freedom and peace are not only intellectual achievements but spiritual and moral achievements as well, demanding a cherishing of the wholeness of human personality and constituting a never-ending challenge to man, emerging from the abyss of meaninglessness and suffering, to be renewed and replenished in the totality of his life. For as one's thinking is, such one becomes, and it is because of this that thinking should be purified and transformed, for were it centered upon truth as it is now upon things perceptible to the senses, "who would not be liberated from his bondage." [1]

There is in mankind today a counterforce to the sterility and danger of a quantitative, anonymous mass culture, a new, if sometimes imperceptible, spiritual sense of convergence toward world unity on the basis of the sacredness of each human person and respect for the plurality of cultures. There is a growing awareness that equality and justice are not to be evaluated in mere numerical terms but that they are proportionate and analogical in their reality.

We stand at the brink of the age of the world in which human life presses forward to actualize new forms. The false separation of man and nature, of time and space, of freedom and security, is acknowledged and we are faced with a new vision of man in his organic unity and of history offering a richness and diversity of quality and majesty of scope hitherto unprecedented. In relating the accumulated wisdom of man's spirit to the new reality of the World Age, in articulating its thought and belief, *World Perspectives* seeks to encourage a

[1] *Maitri Upanishad* 6.34.4, 6.

renaissance of hope in society and of pride in man's decision as to what his destiny will be.

The experience of dread, in the pit of which contemporary man has been plunged through his failure to transcend his existential limits, is the experience of the problem of whether he shall attain to being through the knowledge of himself or shall not, whether he shall annihilate nothingness or whether nothingness shall annihilate him. For he has been forced back to his origins as a result of the atrophy of meaning, and his anabasis may begin once more through his mysterious greatness to re-create his life.

In spite of the infinite obligation of men and in spite of their finite power, in spite of the intransigence of nationalisms, and in spite of spiritual bereavement and moral denigration, beneath the apparent turmoil and upheaval of the present, and out of the transformations of this dynamic period with the unfolding of a world-consciousness, the purpose of *World Perspectives* is to help quicken the unshaken heart of well-rounded truth and interpret the significant elements of the World Age now taking shape out of the core of that un-dimmed continuity of the creative process which restores man to mankind while deepening and enhancing his communion with the universe.

New York, 1954   RUTH NANDA ANSHEN

CREDO

That the resources of imagination and reason are not exhausted.

That a mature humanism requires the understanding which can only be given by an elegant * theory of nature, of organisms, and of ourselves.

That those who are now young may enjoy the experience of watching this understanding develop during their lifetime, if their attention is called to the process now.

That this new insight will arise from a clarified idea of FORM, which the sciences will recognize as the most powerful clue to the relatedness of things within reach.

* "Now, what are the mathematical entities to which we attribute this character of beauty and elegance, which are capable of developing in us a kind of aesthetic emotion? Those whose elements are harmoniously arranged so that the mind can, without effort, take in the whole without neglecting the details. This harmony is at once a satisfaction to our aesthetic requirements, and an assistance to the mind which it supports and guides." Henri Poincaré, *Science and Method*. The belief in the existence of such elegance in nature is the basis of the Credo.

ACCENT ON FORM

I.

The Search for Understanding

Mankind is capable of acquiring understanding but is still abysmally ignorant

WHAT kind of universe is this into which we are born?

Is it the creation of an intelligible God, whose purpose in putting us here we must try to discover?

Or is the universe an assembly of atoms following laws of chance, and mankind an accident?

Or does the universe display an order in which every part is harmoniously related to the whole, if we could but see it?

No one knows the answer to these questions. Nor do we know how much it is possible for man to know.

The mystery of the existence of the universe may forever lie beyond human comprehension. We are part of a cosmic process which cannot be compared with anything else, since it is unique.

Yet it may be useful to pose absolute questions and to consider what kind of answers might be given to them. The human mind has advanced because men dared to ask questions that seemed strange because they pointed toward new problems. No limit can be set today to what the human mind may achieve

1

tomorrow. Speculations on the nature of the universe may prove as fertile as the first attempts of Homo to fabricate a tool.

Moreover no one can live without either conscious convictions or unconscious working assumptions regarding himself and the universe. A latent metaphysic molds every human life. Those who claim they have none still work on rules of some kind. We are all conditioned by our experience.

I have tried to express here my picture of the universe and of man. It is based on scientific knowledge interpreted and adjusted in the light of a personal judgment. For science is incomplete and its present ideas are unlikely to be final.

The first step is to discover what meaning we can give to the question: What kind of universe is this?

We can ask of science: Do such fundamental laws of nature as are already known tell us anything about its basic character? Do the known laws go so deep as to reveal the actual form of the order of nature?

The answer is that we cannot be sure. The known laws are in many respects imperfect and incomplete, and beneath them may lie some deeper and more general order than that which has already been discovered.

So let us put our question in a narrower form and ask: What can science say about the fundamental laws of the three realms of matter, of life, and of mind, or about their relations?

The answer is: nothing *fundamental* yet, on any of these points. There is no satisfactory theory of the fundamental physical particles, of biological organization, or of mental processes, or of their interrelations.

This may be regarded as rather disappointing fifteen generations after the foundation of exact science by Galileo and

Kepler. Many particular facts and partial rules are known, but no truly fundamental principles have yet been discovered, if by that we mean principles which throw a clear light on particles, organisms, or minds, or on their relations. The known laws are probably only special cases of deeper laws still to be identified. On these great issues science is as yet silent, and no living person knows anything for certain. So far we know nothing fundamental about the universe into which we are born.

There are moments when the depth of human ignorance is frightening. We then find ourselves looking into a bottomless abyss and we have to call on our last reserves of courage. What is cancer, that murders the body of a friend; or psychosis, that can destroy the mind; or death, that is for the individual the apparent end to everything? Other moments are less horrible, but perhaps lie even heavier on the conscience and will. When the gravest decisions have to be made, how little knowledge we have of the proper criteria to use! In every realm the deeper our need, the more profound our ignorance proves to be.

I remember two occasions on which this sense of human ignorance came to me with special force, once in a personal and once in a social situation.

The first was when a close friend revealed features which I felt to be nearly psychotic. The friend seemed to be living in a world so far from reality that I was scared. Circumstances prevented me from passing on the responsibility to a psychiatrist. I became occupied with the problem: How did this condition develop? What roles had heredity, parental influence, and personal experience played? I collected what I thought might be the most relevant facts, but I was still unable to gain enlightenment and I fell back on the question: How far could scientific knowledge be of help? From this I reached a further question which startled me: In the entire history of mankind had anyone yet recorded and interpreted the reasonably full story of one human being, showing how all the main physical and mental characteristics had developed?

I realized that if the proper study of mankind is man, this question was of unique importance. But the answer was: No. This had not even been attempted, in fact we had not the knowledge which it would require. There were some rather detailed studies of genius; there were stories of "identical" twins; there were medical and psychological case histories;* and there were some fairly detailed conventional biographies. But what I wanted was not yet feasible. All the biological and human sciences put together did not let us understand even the main features of the story of one human person. The art or science of biography had scarcely begun. For example, we had no idea how a set of genes actually influenced the development of adult anatomy and physiology; nor did we know why one child was generous and courageous and its brother or sister mean and cruel.

So I could achieve neither intuitive nor scientific understanding of my friend. . . . But it did not matter. By divine fiat, good luck, or organic vitality—how can I claim to know which?—my friend's life grew richer and the frightening symptoms passed away.

The other occasion was in London during the winter of 1941–42. Europe had fallen; the most hideous mania in the history of the West had seized power; the outlook for Britain

* Perhaps the psychobiological studies of the American psychologist Adolph Meyer come nearest to what I was looking for.

was grim; the spiritual values of Western civilization had for long been declining and then in wartime were nearly forgotten.

I asked myself, was there any rhyme or reason in this? Was it indeed the decline of the West? Was there any philosophy of history or any scientific interpretation of the biological and social development of man that could throw light on these crises of civilizations? No, there was not. Toynbee, for example, neglected the biological background of human development and the profound impact of science. The human race was here even more ignorant than in relation to physical and mental disease. No man had ever lived who could estimate the significance of his own culture in the over-all story of mankind. One could not even hope for any assurance in a period so short as one lifetime. . . . This situation led to no happy ending; the frightening symptoms are still present.

On both of these ocasions I had felt the vast hopelessness of human ignorance. And yet my mind rebelled at the idea that anything that could happen was completely beyond comprehension. I never doubted that there existed an order of nature which included man and was progressively accessible to human intelligence. It followed that all human distress occurred within the natural order and must in some manner be transformed, though not necessarily removed, by a deeper understanding of that order. The point was to discover in any situation the *optimum line of development,* even if that was only to relax and to seem to do nothing. What mattered was to reduce the unnecessary frustrations and to enrich life, even if that could only be done by accepting the inescapability in every individual life of bitter experience.

Later I came to realize that it was no good taking particular historical situations too seriously. It was ridiculously early to condemn man as a failure. Homo sapiens was the kind of species that could not live properly in accordance with its capacities without a way of thinking appropriate to the particular stage of development which it had reached at any time. The present attitude of the West was the result of a long development through the Renaissance, the Reformation, the Enlightenment, and the growth of science and technology, and the modern world had not yet found a system of convictions appropriate to an age of science. A naïve reliance on an incomplete and unbalanced science was no substitute for what Christianity had meant to an earlier time. To the biologist or anthropologist the malaise of the twentieth century was in no way surprising. Homo sapiens was a biologically immature species that had not yet fully developed the latent capacities of its brain, and might be going through a crisis which was an inescapable phase of its social development. We should try to understand rather than to condemn without proper trial.

If one is able to accept this long view, it is human understanding rather than human ignorance which is astonishing. The remarkable fact is not that no man has ever lived who has possessed the kind of fundamental knowledge we so badly need, but that the race has managed to achieve any understanding at all, and how much it has gained.

Here the facts are certainly extraordinary. The interbreeding species Homo sapiens, marked by an average hereditary equipment similar to our own and by characteristic social habits, emerged from its ancestral Homo stock between four hundred and one hundred thousand years ago, and may only have developed the faculty for articulated speech around fifty thousand years back. Making allowance for prior types of Homo that disappeared we can regard it as certain that *neither five hundred thousand years ago, nor at any earlier time, was there either on the earth or elsewhere in the solar system any organic species possessing any knowledge based on the units of thought which we call ideas.*

At that time there was no articulated speech, or spoken sounds broken up into meaningful units; no thought communicated by dividing it into distinct and significant parts and separating it from immediate organic situations; no systematic conceptual understanding. Many highly complex systems of symbolic communication were already in use, for example in the dance of bees and the song of birds. But these did not involve units of thought detached from immediate activities. There was as yet no germ of intellectual understanding of the relatedness of things in an objective world. The potentiality for this was already present, say four hundred thousand years ago, in the hereditary equipment of the emerging species Homo sapiens, but what was to prove the outstanding characteristic of mankind had not yet been manifested.

Then very slowly, though perhaps in minor rushes and relapses, starting from this absolute ignorance and without the guidance of any conscious aim, the human species began its incomparable adventure of discovery and self-discovery. Discovery of external nature went hand in hand with the discovery of human faculties, the arts of action developing in parallel with the art of thought, the entire process resulting in the progressive maturing of capacities which had been latent in the human hereditary make-up. From prehuman gabbling, gesture, and dance there developed—perhaps starting rather abruptly a hundred or fifty thousand years ago—first language, then script, and finally the culture that blossoms in Newton and Beethoven.

This is no less than a miracle. For this cumulative process of the realization of human potentialities took place without conscious design, as part of the pervasive continuity of natural process. In some sense which has still to be made clear natural law guided the development of man. This flowering of latent capacities must have been implicit in the condition of man all along, though it required a favorable environment to bring it out. We may consider this a miracle, but it simply proves the richness of the "unconscious knowledge" and biological assets of the species, or of the laws which governed the process.

I shall call the essence of this characteristically human process which works mainly cumulatively and unconsciously, *discovery*, including under that term the simultaneous and interacting aspects of the external discovery of nature and the internal discovery of the latent capacities of human nature. All creation, imagination, and invention are aspects of this dis-covering of the new.

The historical fact of human discovery is more astonishing and more significant than human ignorance. For without the fact and the experience of past discovery we could not even have the idea of ignorance. Our awareness of our ignorance is evidence of our faculty for discovery. The abyss is not bottomless, we have made a start.

Even when human ignorance was still absolute there was present in organic nature a formative process, a surplus vitality, a creative, exploratory, or inventive instinct which, when the time came, would shape in human brains and minds ideas that would bring enlightenment. This organic faculty for achieving understanding from ignorance is the one unchallengeably favorable fact about man. He can grow in understanding.

This fact is neither trite nor trivial. None of the world

religions has adequately recognized the supreme importance of this human faculty for progressive discovery. Science does, and on this account alone can speak with authority when these long issues of the human past on this planet are in question. In his ability to grow in understanding man seems to touch the divine, and yet here it is unconscious processes which provide the foundation, and science which takes up the service and protection of this great faculty. What irony that the unconscious and science should be the servants of the noblest element in man!

We know very little about the unconscious mental processes which provide the basis for the creative, imaginative, and inventive faculties. Like many other organic processes they work, and perhaps even work best, without our knowing how. But it is clear that no arbitrary accidents or merely occasional *tours de force* can account for the inexorable continuity which, transcending all rhythms and setbacks, has led from the earliest members of the species to where we are now. And since the individuals who achieved the myriad steps knew little or nothing about what they were doing, the main credit must go to unconscious processes.

In retrospect we can observe that in its scarcely conscious search for understanding the species invented for itself one supremely valuable kind of instrument: *ideas*. Their value was not discovered as the result of any deliberate investigation. Ideas arose as a residue of barely conscious mental processes, doubtless associated with processes occurring in the brain, but not yet understood. All we can say is that some kind of formative process in the brain shapes new forms of activity and response within the plastic records of past experience, and that in certain circumstances these new unified forms or clarified patterns become the object of our attention, that is, become conscious as "ideas."

This view of ideas implies that man makes them for himself, though by methods of which he is unaware. Plato's conception of eternal Ideas or universal intelligible Forms was different. He imagined that the Demiurge, or Skilled Workman who made our world, desiring that all created things should be as like Himself as possible, took from His own real world the Models (or Forms, or Ideals) representing the generic idea of everything, and used them to make Copies in the world of appearances. Thus the transient world of phenomena came into existence.

Plato's image has been of great importance for the human mind, for it contained the fertile thought that within every changing appearance there lies an unchanging factor, in medieval language the form that makes every particular thing what it is. But today we who accept the outlook of science enjoy an even more powerful image: we see the creative process not in some divine act in the past, but in the continual daily working of our own minds, even the humblest. For every person over one or two years of age and not mentally defective is perpetually shaping new patterns of thought, new little ideas or hunches, every day of his life. We have brought the formative process back into nature and into ourselves, and here we are wiser than Plato. Mankind forges its own instruments for the voyage of discovery.

But the major instruments, the new *primary* ideas, take a long time to perfect, much longer than an Egyptian pyramid for example. The largest pyramids may have taken some twenty years to build, but *in the two thousand five hundred years or hundred generations of Western thought only some ten or twenty primary ideas have been produced!* I am here leaving aside ideas related to subjective experience such as God, Beauty, Goodness, and Justice, and considering only those ideas which serve directly as instruments for understanding the universe. Here are some of those which seem to me most important. Where possible I have added the name of a person who either gave precision to the idea or greatly extended its use.

NUMBER	Pythagoras
SPACE	Euclid
TIME	
ATOMS	Democritus
ENERGY	
ORGANISM	Aristotle
MIND	
UNCONSCIOUS MIND	Freud
HISTORICAL PROCESS	
STATISTICS	
(FORM)	
(STRUCTURE)	

These twelve ideas can be regarded as covering all the primary insights which the Western mind, and that means human systematic reason, has yet had into the nature of the universe. These are the main instruments of intellectual understanding which the race possesses today. Each has gradually grown clearer and more definite and has been stabilized by persistent use. If we had to send a summary of scientific knowledge by radio to a distant star these would contain the nucleus of the most reliable information.

You will notice that two are put in brackets. The other ten

are all fairly clear ideas, clear enough at any rate to have university chairs allotted to them (if there is a Professor of the Unconscious). But Form is still an ambiguous and fertile conception, capable of meaning almost anything. It is pregnant with untold possibilities, for confusion if mishandled, and for new clarity when we can find the way. Form is the dark horse. And Structure—well, more of that later.

Some of these ideas are very old. Number, Space, Time, Atoms, Organism, and Mind had clear meanings for the ancient Greeks. The Hebrews in the centuries before Christ had a vivid conception of Historical Process. Energy grew precise around 1850. Unconscious Mind was much in the air around 1870. And Statistics, often held to be the darling child of this century, is a wiry old man who has worked hard for over a hundred years; in 1834 there was a Statistical Society of London. These ten ideas have acquired a new look recently, but they are all thoroughly respectable, and that means at least fifty years old.*

In one sense Form is also as old as any of them. The Greeks had the idea, the Middle Ages gave it a new significance, and hardly a century has failed to enjoy its own conception of the meaning of Form. We shall be glancing at some of these in a moment. But our own time has not yet done this. To be more exact, the twentieth century has not yet given to the concept of Form its own standard of precision, and it may be necessary to do this before other ideas such as Organism, Mind, and Unconscious Mind can also be made as precise as we would like.

However the history of thought does not follow a logically

* When a new idea is widely recognized as important its most productive phase is already over.

necessary sequence which can be recognized at the time. Ideas are taken up because for some subtle reason they appear to be timely or interesting, and that means to the young speculative minds of each period. What counts today is: What kind of ideas are likely to appeal to the generation of imaginative thinkers who are now, say, fifteen to thirty-five years of age?

I hazard a guess that the kind of young mind which is likely to be most productive in the realms of science and philosophy during the next few decades will be drawn to ideas which promise:

To bring a new unity and order into specialized knowledge.

To express change and process, rather than permanence.

To reveal a single process, underlying the partly verbal dualism of matter and mind.

To throw light on the relation of part to whole in all realms.

This is my own guess, but it represents the common ground, not of a "school of thought," but of countless individual workers in many branches of philosophy and science. Twentieth-century man is in great need of a new scientific "myth of creation," a cosmology or way of thinking satisfying these conditions and appropriate to the present state of knowledge.

No one can reliably foretell the future. But I believe that the idea of Form, suitably clarified and strengthened, will go far to achieve these aims and will transform many pressing problems, scientific, philosophical, aesthetic, and moral. This book is written as a philosophical speculation which may help to prepare the way for the science of tomorrow. In exact science one constructive achievement is worth a world of dreams, but dreams often have to come first.

II.

What Is Form?

*Men have long been seeking an adequate idea
of form*

THE pedant may say: "form" is whatever we define it to
mean, but that is not helpful. What has "form" meant in the
past, and what is the best meaning we can give to it today?

A complete answer would amount to a history of thought,
for in one sense everything possesses form. In some contexts
the Greek words *Eidos, Schema,* and *Morphe,* and the Latin
word *Forma,* which are often translated as "form," mean no
less than "the qualities which make any thing what it is." If
we accept this meaning all philosophy and science can be
regarded as the endeavor to study the forms of things and to
discover the underlying formative principle which brings all
things into existence and makes them what they are. Every
period in history and every school of thought has had its own
idea of this supreme formative principle, though it was seldom
called that. Here we can only glance at some of the more
important.

The scripts which tell of the first civilizations and studies of
the most primitive cultures still in existence show that at

an early stage in its development human consciousness is dominated by a pervasive sense of process, though to give it this name is to describe in abstract terms a condition which is naïve and unreflecting. The primitive mind does not consider its own workings, but is spontaneously attentive to the perpetual transformations of nature and life. Thus we find that among the earliest interpretative principles which man invented were *myths of process and transformation*. It is probable that all the primitive cultures possessed some myths of this character which were unreflectingly accepted as the explanation of experience, the reason why things were as they were.

During this early phase of human development existence was viewed as a process of dissolution and creation, evidenced by death and reproduction, mortality and fertility, decay and growth. A perpetual process of transformation was recognized in human life, in organic nature, in the alternation of the seasons, in the waxing and waning of the moon and the rhythms of the sky, and this transformation was sometimes traced to its source in myths of the gods that die and are reborn, Osiris and Adonis.

Beside this concrete and dramatic sense of experienced transformation there existed an awareness of order and regularity, sometimes expressed in the recognition of cosmic powers measuring the rhythms of time, such as Sîn, the Chaldean moon-god. The man who today watches the nightly changing form of the moon and the woman who marks her calendar are following what may well be the oldest of human traditions, for the alternation of the seasons, the moon rhythm, and the physiological cycle in woman have been matters of the highest importance from the times when man first began to distinguish one day from another. The moon-god, symbolizing the organic rhythm of growth and decay, was one of the first formative principles to be conceived by the human mind.

The thought of the earliest civilizations was concrete and pluralistic, recognizing many ultimate principles, and it was probably not until the time of the Greeks that an attempt was made to find a single principle which might account for everything. We need only take one early example. Thales found in water the basic principle, the source and foundation of all existence. Here the important feature is not the specific assertion but the daring attempt to discover one principle that underlies everything. Such attempts can have no prior rational justification, but we may assume that they express a need of the human organism to seek order when confronted with embarrassing complexity. The intense need becomes transformed into an intellectual conviction, a belief that an order exists which can be found. Skepticism can challenge the basis of that belief, but experience has proved its efficacy in promoting discovery.

The vast proliferation of knowledge and the lack of adequate simplifying ideas is a grave embarrassment today. In this situation it may be wise to aim high, and to start by looking for one principle that can throw light on everything. I therefore dedicate this essay, with a vivid sense of its inadequacy, to the Unknown Formative Principle, the Logical Structure of Becoming, the Logos of Chronos, which will be known to future generations and by which they will think and live.

The search for a universal formative principle has already occupied many centuries, possibly because it has been prejudiced by the human fear of change. Like the human infant man displayed his preferences almost at the start, setting his heart on symbols of permanence, symmetries whose perfection sought to deny change. The Egyptian and Mexican pyramids are evidence of this early linking of symmetry with stability and permanence, and of the deep attraction for the human mind of a dream in which time is overcome. And here at the very beginning we meet the ambiguity of every powerful emotional symbol: for symmetry, being static, is a symbol of death, not of life. The pyramid designed to protect the dead Pharaoh and his victuals so that he might survive and conquer time becomes the most deathlike of tombs. What is less like life than a pyramid, and less human than a perfectly symmetrical face or handwriting? Man is attracted by symmetry, and yet symmetry is a dead end.

The Egyptian artist saw the pattern of human life and its natural setting mainly in a flat geometrical frame, perhaps echoing the plain of the Nile bed. Egyptian architectural decoration and art were mainly concerned with the plane; there is hardly any evidence of a reflective awareness of the third dimension of space. It is remarkable that Egyptian friezes and patterns exploited all the possible types of two-dimensional symmetry at a time when there was apparently no trace of awareness of the possibility of representing three-dimensional visual perspective in a plane picture. Egyptian life, at its best, was rich in imagination, poetry, affection, and good living, but no Egyptian, it seems, held in front of his eyes a strip of material with a rectangular hole cut in it in order to see the world as a picture with perspective, as did Leonardo da Vinci.

Nine hundred years after Akhenaton, who marks the end of the great dynasties, Pythagoras comes from Greece to Egypt, learns all he can, and then in Southern Italy founds his unique brotherhood, in which religious mysticism and social idealism were inseparably blended with intellectual enthusiasm for the vision of Number as reality itself.

Other periods would find no difficulty in understanding this single passion and its manifold expressions. Certainly Kepler would not, whose religious devotion expressed itself naturally in numerical research. Nor surely would Einstein whose life work displays just this single devotion. But such unity of purpose seems to be beyond the understanding of so disintegrated a time as ours. For example, the *Encyclopaedia Britannica* (14th Edition) says, "The scientific doctrines of the Pythagorean school have no apparent connection with the religious mysticism of the society or with other rules of living."

There is a nice blind spot! The Pythagoreans were drunk with the vision of God in Number, like many after them, and scorned more sensual pleasures. That is an orthodox way of putting it, which would be acceptable to all true Pythagoreans down the centuries.

But there is a deeper interpretation of the Pythagorean ecstasy. The Pythagorean community holds a unique place in the history of thought because here for the first time the human mind was fascinated by the music of its own thoughts. Moreover this bliss of self-contemplation was not the lonely introspection of a single mind but the collective enthusiasm of a group sharing one emotion and establishing a continuing tradition of their form of worship.

This bears repetition. In the Pythagorean school the human mind was first intoxicated by the harmonies of its own processes. Like a girl suddenly aware for the first time of the grace of her own body and permanently changed by that knowledge, the Pythagorean was so delighted by his own thoughts that he made an irrevocable mark on the history of the species. If we neglect the social aspect of their doctrines, we can picture the mood of the brotherhood as that of an enchanted garden where everything is number, harmony, and proportion. The magic of this narcissism is all the greater because the devotee believes that what is from one point of view an image in his mind is actually the real world. The threat of time and the frustrations of life are forgotten in this retreat. Here nothing is created or destroyed, nothing truly comes into being or fades away, for everything exists by its own numerical laws and partakes in the eternal music of the spheres. The storms of nature have been subdued to play a pastoral melody on a Greek reed.

Every vista in this Pythagorean garden is an avenue of perfection, a radiance of number or form revealing the true nature of things, for number is the limit which gives form to the unlimited stuff. The numerical form of every thing is its characteristic principle, and musical harmony the very voice of nature. Number, perfect and unchanging, holds the secret of every thing, physical, aesthetic, and moral. The true worship implies renunciation of the flesh and provides the strength to accept the unfortunate mischance that the Pythagorean mind is enchained in a corruptible body.

For those who lack the organ for mathematics, or were robbed of it by mishandling, no words can replace the experience of aesthetic joy in the symmetry, necessity, and unexpectedness of mathematical reasoning. It is no exaggeration of this experience to say that here it seems as if reality had become blissful and bliss real. For the unexpectedness of a mathematical result gives us the feeling that it is not our own creation, that the world of number exists of its own right, while its necessity and symmetry are balm after the ragged edges of life, or pure joy to those who do not yet know them. The appeal of mathematical form reaches deep into human character.

If this be true, small wonder that the Pythagorean vision, handed down or rediscovered, has fascinated so many of the greatest. For great minds are those with a great capacity for fascination and enthusiasm. Plato, Archimedes, Euclid, Kepler, Newton, Maxwell, Einstein—all these surrendered to its magic and were fertilized by it.

What a field for study lies in the relation of the religious to the mathematical passion! Consider Kepler striving against every difficulty for twenty years to reveal the simplicity and beauty of the divine mind in the motions of the planets. Or Newton developing his incomparable intuition of a timeless geometry controlling the system of the world in order to justify the works of God to man. Even those who acknowledge no one privileged religion or sectarian monopoly of human wisdom must ask themselves whether the burden of the endeavor to discover a single natural order can be borne without the support of an emotion and a conviction that can only be called religious. Why endure the strain of the search for unity unless one believes that it is there?

Number seems to be at once real and aesthetically satisfying; hence its appeal to this temperament. Certainly no tradition in any culture has so powerfully inspired the search for precision, clarity, and simplicity within the complex and confused appearance of the sensible world. The siren call of the Pythagorean melody is the greatest temptation offered to noble spirits: the promise of harmony, if the burdens of the personal life are neglected.

> There's not the smallest orb which thou behold'st
> But in his motion like an angel sings,

Still quiring to the young-ey'd cherubins;
Such harmony is in immortal souls. . . .
—*The Merchant of Venice.* Act V Scene 1

Such Pythagorean harmony is ours—in so far as we manage
to live in eternity and to neglect the burdens of time.

For some the Pythagorean emotion is sacred and can
support a lifetime. But all are not made or allowed to live
in that pure dimension. For others the vision fades when they
realize, as young lovers may after the first joy of mutual discovery, that the adored object was to an unknown degree the
creature of their own imagination. She is really quite different,
and so is the real world of natural process, of natural history,
and of men and women.

For if a God created this universe in His own image, He
certainly was not a Pythagorean. Once we recognize this fact,
the garden loses its magic and begins to look like a museum.
As we glance around we see how narrow the selection has
been: there are the five Platonic regular solids; the circular or
elliptical paths of the planets; a few geometrical models; some
vibrating strings; patterns made out of the integers: 1, 2, 3,
4 . . .; and not very much more. There is no butterfly or fish
or scorpion; no nude; nothing organic, subject to reproduction, growth, and decay. The passions of birth, of love, and of
death find scarcely an echo here; it is the cloister of a community essentially masculine and monastic, devoted to spiritual
delight, and welcoming as members only those women who
felt and thought and acted like Pythagorean men.

The Pythagorean cosmology is perhaps best described in
parts of Plato's *Timaeus.* For contrast, consider Leonardo
meditating on the possibility of regarding the earth as an
organism:

Nothing originates in a spot where there is no sentient, vegetable, and rational life; feathers grow upon birds and are changed every year; hairs grow upon animals and are changed every year, excepting some parts, like the hairs of the beards of lions, cats, and their like. The grass grows in the fields and the leaves on the trees, and every year they are in great part renewed. So that we might say that the earth has a spirit of growth; that its flesh is the soil, its bones the arrangement and connection of the rocks of which the mountains are composed, its cartilage the tufa, and its blood the springs of water. The pool of blood which lies around the heart is the ocean, and its breathing, and the increase and decrease of the blood in the pulses, is represented in the earth by the flow and ebb of the sea; and the heat of the spirit of the world is the fire which pervades the earth, and the seat of the vegetative soul is in the fires, which in many parts of the earth find vent in baths and mines of sulphur, and in volcanoes, and at Mount Etna in Sicily, and in many other places.

In the absence of fundamental knowledge on the relation of inorganic to organic phenomena, even the greatest minds suffer impediment. For Leonardo here shows little advance on Aristotle, and Whitehead, a few hundred years later, displayed his and our ignorance in finding himself compelled to call the atom an organism. Where knowledge is lacking courageous minds give play to their fancy, even though it may sometimes blur our vision of the truth.

With Leonardo we are back in the world of time. In place of a static perfection abstracted from the world of appearances, the emphasis is on the rhythms of the visible changes of the organic world and the parallels which they suggest. Mathematical harmony has given way to natural observation and analogy, and the recognition of process has brought with it an inevitable loss of precision.

Yet if we examine Leonardo more closely we discover that even as an artist he is often very systematic. We find him reminding himself to consider all the positions of the human body: "repose, movement, running, standing, supported, sitting, leaning, kneeling, lying down, suspended, carrying or being carried, thrusting, pulling, striking, being struck, pressing down and lifting up." There are, he tells himself, ten types of noses in profile, and eleven in full face.

And when, following Brunelleschi, he had discovered how best to express the laws of visual perspective, by drawing straight lines radiating from the human eye to all visible objects, he breaks out:

> Here (in the eye) forms, here colors, here the character of every part of the universe are concentrated to a point; and that point is so marvelous a thing. . . . O! Marvelous, O Stupendous Necessity—by thy laws thou doest compel every effect to be the direct result of its cause, by the shortest path. These, indeed, are miracles. . . .

Each age expresses its state of awareness in its own sense of form, as evidenced in the arts or in philosophy and science. The Egyptian flatland was succeeded by the Greek sense of balance, proportion, and symmetry, and Aristotle developed an awareness of organic forms which had been lacking in the Pythagorean school. Then with the rise of Christianity a new aspiration entered the human heart, a burning desire to transcend the limitations of the transient world and to reach, like Pythagoras and Plato, a Universal Form which could satisfy the human need to pay reverence to something nobler and more lasting than oneself.

The Greek sense of human self-sufficiency faded, and in seeking the infinite men began to create a world of thought, a scholastic discipline of reasoning, to compensate the inadequacies of the individual. Thus in the Middle Ages form was interpreted, following Plato and Aristotle, not merely as a visually perceived shape, but as an internal principle of being, and the world was viewed as a divinely ordered hierarchy of forms. Around 1250 we find Thomas Aquinas regarding *forma* as the essential quality or determining principle of every individual thing. Here form has become so abstract as to possess only a logical or potential meaning, for we still have to discover what is that "very essence."

To Pythagoras that essence was Number; to Plato the Eternal Forms; to Aristotle a striving toward Realized Form; to Euclid the quantitative relations of space; to Aquinas the mind of God; to Leonardo and Francis Bacon the arrangement of the spatial parts that make up a whole. And in the title of this book form means spatial shape including internal structure.

Thus through the centuries the word "form" (*forma*) has continually undergone changes of meaning, marking at each stage not a definite discovery, but a program or a declaration of faith. Men were searching, though perhaps unconsciously, for an adequate idea of form.

Between 1600 and 1650 a dramatic transformation occurred. Kepler and Galileo proclaimed the doctrine that measurement holds the clue to the understanding of nature, and Descartes recognized the importance of exact analysis. The impact of this fresh development of an old idea was unprecedented in its violence. From 1650 onward form, in the sense of spatial shape and contour, has taken a secondary place.

For the contemporary scientific mind the real world is built up of minute parts; the laws determine how the single ultimate parts behave; and form—far from being the essence—has become a relatively trivial consequence of the behavior of the parts.

The change-over was remarkably sudden. Around 1600 Francis Bacon wrote: "the form of a thing is its very essence." By 1647 "formal" had come to mean "*merely* a matter of form," and whatever was trivial could be called a (mere) formality.* Fifty years had turned the dominant attitude upside down; what had for one generation been all-important had become for their grandchildren the least so.

Efforts were made to resist the dominance of the new analytical science. Throughout the eighteenth century the leaders of the developing science of biology were studying the anatomy and external forms of organisms and arranging the species accordingly. Around 1800 the term "morphology" was in use, and sixty years later Darwin provided the first comprehensive theory of the evolutionary origin of organic forms.

But the most impressive achievements of science lay elsewhere. Using the methods of empirical and mathematical analysis sanctioned by Galileo, Kepler, and Descartes, Newton, Faraday, and Maxwell made their incomparable discoveries. In a sense which Pythagoras can never have anticipated nature seemed to have justified him: once again God appeared to be a mathematician, and a very accurate one. For the God of the

---

* I consider that the time has come to repudiate this Cartesian reversal of meaning. Here and elsewhere I shall use *formal* to mean "concerned with spatial form" (as contrasted with individual constituent parts). The theme of this essay is that the next major advance in science will depend on the use of formal principles, in this restored sense. The word "abstract" can be used in place of "formal" to mean "concerned only with mathematical or logical structure."

mathematical physicists enabled them to achieve unchallengeable advances; in comparison all other ways of thinking seemed to be sterile.

So arrogantly successful a doctrine inevitably produced its rebellions. From 1800 onward the vitalists began to claim that the unique properties of organisms were evidence of a principle which could not be covered by mechanical or chemical explanations. And later toward the end of the nineteenth century, psychologists concerned with perception were emphasizing the importance of system or pattern characteristics, properties of the whole rather than of the parts. By 1920 this movement became known as the Gestalt school and it exerted considerable influence, but its principles lacked the clarity and precision of the analytical method.

In the meantime another idea had begun to emerge, more powerful than any since Newton, a revolutionary intellectual force which must sooner or later transform human thought in ways which we cannot now foresee. As I have said, a new major idea is never recognized in its full power for at least fifty years after its first formulation. Marxism was launched in 1848, and the Russian Revolution took place in 1917. Planck discovered the quantum in 1900; we do not fully understand it yet.* The idea which concerns us now first began to grow clear around 1910, and it will only attain its full power during the next generation. It can be expressed in one word: STRUCTURE.

* Here is one example of something which is not understood. The number 137.0365 . . . appears in experiments on the interactions of light and electrons, but there is no acceptable explanation of it. The mind rebels at the idea that nature contains arbitrary and meaningless numbers, and prefers to believe that we have put this number into our picture of nature by using inappropriate ideas. If we could see deeper and improve our ideas we might understand why the number has just this value. Eddington produced an explanation, but it is hard to find anyone who accepts it.

This idea, "structure," stems from much earlier roots, but it is really a twentieth-century invention, indeed it is *the* creation of this century as far as human understanding is concerned. It will be remembered when many of the noisier achievements of our age are forgotten.

"Structure" is a name for the effective pattern of relationships in any situation. This may sound abstract, but the thing it stands for is not so. Take father-mother-child for example. That is a most effective pattern of relationships. It is a triad with asymmetrical relations, each subject to subtle changes. It is always set in an environment; it has inherent developmental and explosive tendencies, attractions, and repulsions, and unpredictable potentialities. A searching study of the variants of that triadic structure would cover a large part of human knowledge concerning man.

But here we are concerned chiefly with physical structure and with what the physicist calls *ultimate structure*. By 1910 countless experiments had "established the reality of atoms." More precisely, the ideas of atomism had proved so successful in interpreting physical observations that it was almost impossible for a working physicist to think in any other manner. Material objects were undoubtedly granular in constitution, and were best thought of as composed of vast numbers of tiny atoms or particles. This is the first step toward the idea of physical structure: *Everything is composed of ultimate particles.*

These particles are subject to very stringent rules of behavior. Only under exceptional conditions (in gases at very high temperatures or low densities) do they move at random, as was wrongly assumed in the dominant theories of the nineteenth century. On the contrary the typical condition is one of rather high regularity, symmetry, or ordering, as for example in crystalline materials, in organisms, in plastics, and in liquids—for even in liquids every small region is highly ordered. Though some prejudices remain from the last century the facts are unmistakable, and this is the second step toward the idea of structure: *In general the particles display a high degree of ordering in space; they tend to assume regular or ordered arrangements.**

This means that the normal states of physical systems display definite spatial patterns such as the linear arrays of atoms and molecules in a crystal, the chains of atoms in fibers, and the still obscure but highly significant arrangements in genes, viruses, and all working parts of organisms. Here it is the pattern or arrangement which counts for most purposes; the individual particles are indistinguishable and may come and go at random. Indeed apparently the only function of the particles is to build up the patterns, for it is these latter that we actually observe. This is the final step: *The "form," in the new sense of the underlying structural pattern, is more important than its material components, which lack individuality.*

Thus the twentieth-century idea of structure amounts to this. If one magnifies anything enough one reaches a characteristic structural pattern which is fundamental for the understanding of the properties of the thing. In every situation it is the ultimate structural pattern, rather than the individual material constituents and their supposed properties, which matters. The implication is that to understand anything one must penetrate sufficiently deeply toward this ultimate pattern. It is almost as though the pattern determined the properties of its constituents, rather than the other way round. Complete under-

* Requiring relatively few numbers to represent them.

standing would include the nucleus of the atom, the atomic structure, the molecular structure, and (in an organism) the biochemical, physiological, organic, and perhaps even the social structure. That we cannot yet achieve, but the structurally minded scientist always keeps this complete hierarchy of structure in mind, from the atomic nucleus up to the organism—or whatever else he is studying.

Thus structure is spatial form seen in its full complexity. This conception greatly extends the ancient, medieval, Renaissance, and pre-twentieth century idea of spatial form. Though Newton had a dim idea of the hierarchy of structure, to an eighteenth- or nineteenth-century biologist the form of an organism was its external shape together with just as much or as little of its anatomy and physiology as he cared to include. But tomorrow no scientist will be able to consider the external shape of anything without simultaneously studying the inner structure which gives each thing the properties it has, and this may include even the ultimate nuclear and atomic structure.

It is evident that we are coming round full circle to the medieval conception of form as that which makes a thing what it is. Yet strictly it is not a circle, but rather a circuit of an advancing spiral, since form is now grounded in ultimate structure.

There are many scholarly books with illustrations telling the reader what is known about the nebulae, the night picture of the sky, the shape of the earth, and the forms of organisms, of crystals, molecules, and so on. Many such specialized works are available, and there would be little point in trying to do here in brief what is done in such detail elsewhere.

The aim of *World Perspectives* is to give perspectives of the major issues of the second half of the century, and I shall therefore concentrate on what is *not* yet known, but perhaps will be in the coming years. This essay is an outline of what I believe will shortly be achieved by a new science of form, rather than a survey of what has already been established by existing sciences. It is a speculative anticipation of the way in which the new idea of form may reveal the relationships of the specialized branches of knowledge and so deepen our children's and grandchildren's understanding.

Since this book is an anticipation of the future cast in the form of a personal affirmation, it seems appropriate that it should be given positive expression, without continual qualifications. No claim of scientific authority is implied, and it will not be surprising if some of these suggestions prove mistaken.

Every attempt to help to restore unity to knowledge and immediacy to thought can claim the sympathetic consideration of all who are aware of the dangers of the present situation, however speculative the attempt may appear or challenging to current assumptions. Blemishes of expression will be forgiven by those who know how difficult a task awaits anyone who, while honoring the specialist disciplines, seeks a simpler and more comprehensive picture.

Understanding means rational insight into the simple relationships between things. We shall be concerned here mainly with intellectual insights into the order of nature. Literary and artistic form and the aesthetic and moral aspects of experience will not concern us directly.

But—and it is a far-reaching "but"—I believe that each of us is a changing form in a universe of forms, and that no perceived form can be emotionally neutral to us, if we let ourselves become aware of our latent responses. At bottom we cannot be indifferent to anything. Everything in this universe

bears some relation to our own nature, its needs and potentialities. Every process mirrors some process in ourselves and
evokes some emotion, though we may not be aware of it.

Symmetry attracts and repels, for it is at once perfection—
and death. Mathematics attracts some and repels others, for it
symbolizes the two-fold ability of reason: to construct reliable
systems of reasoning and to evade problems where one's own
personality is involved. A growing plant attracts, and so does
a ripe cornfield. A raging tiger is fine, but frightening. The
starry night moves us. Wherever there is interplay of order
and disorder, of balance and polarity, of uniformity and contrast, we become aware of an echo inside.

These stirrings are evidence of our oneness with the universe
of form, in no sentimental sense. The universe is not favorable
or unfavorable, but just as ambiguous as is human life itself.

One source of this ambiguity is the duality of order and
disorder. There is a tendency toward order, in nature and in
ourselves, and our first naïve impulse is to welcome order and
symmetry. But then we realize that where all is order and
symmetry there can be no life; the tendency toward symmetry,
if interpreted as movement toward an ideal, reveals itself as
the death wish.

Life is the ordering of disorder.* The ambiguity arises when
in Sunday School fashion we try to label order as good, and
disorder as bad.

There is less need here explicitly to consider the aesthetic or
the moral sense, because they are both expressions of this basic
principle of life: the tendency toward the ordering of disorder.

* This assertion lacks scientific precision, for the concepts of "order"
and of "disorder" require clarification. A step in this direction is made
in Chapter VII.

III.

The World of Form and Structure

*New vistas of form and structure will open out
for the next generations*

HENRY JAMES says of one of his characters, "the love of knowledge coexisted in her mind with the finest capacity for ignorance." She shrank from probing the unlighted corners of human nature, respecting the privacy of others.

Unfortunately the scientific specialist sometimes shares this capacity where it is less respectable; in teaching his pupils he forgets to mention the unlighted corners. I cannot think of a university textbook, or survey of any established branch of knowledge * intended for students, which devotes a chapter to an outline of the most important matters on which the author and his colleagues are ignorant. The advance of medicine, for example, would be appreciably assisted if the student were

---

* Excluding new subjects like Cosmology, where there is no established theory and only reasoned speculation. . . . Since writing the above I have discovered *Prospects in Psychiatric Research*. Edited J. M. Tanner, Oxford, 1953. This is the report of a conference expressly devoted to the question: "What are the ignorances which today principally hamper our understanding of the nature, prevention, and cure of mental illness?" A splendid example for pure science.

instructed in his final year regarding some of the clinical problems on which more information is urgently required. Good teaching in any realm should refer to the regions where the growing points of knowledge are invading the unknown. There appears to be a conspiracy of silence about this widespread failing.

If this is so in medicine and other established specialisms, how much more so in respect of form! For there are no academic chairs, no degrees, no subsidies, no general textbooks, no marks of recognition to promote the general study of form. Only a few great names: Pythagoras, Aristotle, Leonardo, Goethe, d'Arcy Thompson, some lesser lights, and a scattering of individuals who find their way to each other by their common interest, have kept the general theme alive.

If there is no comprehensive textbook, no encyclopedia on what is known regarding form, it is scarcely surprising that no study is available on what is not known, but should be soon. Since little basic knowledge of the laws of form has yet been gained, all that is possible today is the outlining of a program, the provision of some hints regarding what may be discovered soon. Nothing would more effectively stimulate young minds to fertile endeavor than a searchlight on these dark corners— let the great Foundations take note. By Pythagoras and Leonardo, is not form worth one thousandth of what flows to the nucleus and the atom? We are indeed a blind race, and the next generation, blind to its own blindness, will be amazed at ours.

So these pages offer some hints on what will probably be known to the reader's children or grandchildren. If a few of these are right father or mother can be ahead of their children by reading these pages, or can at least try to be. And that attempt will be useful, for we are here concerned not with established scholarship, but with heightening the sense of life and promoting the adventure of discovery.

I am certain that an imaginative parent or teacher, catching the next generation at the right age, could enrich the lives of some at least by dropping hints of the magic of form. For that is nothing other than planting in the young mind the sense of the incomparable mystery of the universe being as it is, a cosmos of forms appearing, transforming, and disappearing in accordance with laws of which no human being has yet had an inkling, for even today there is no exact science of changing forms. Is it adventure or fame which the young spirit desires? Allow fifty years to pass and there may here be greater surprises than the space aeronaut will achieve in five hundred.*

Grant fifty years, and with them your most sympathetic imagination. Be fifteen, seventeen, nineteen once again, innocent of all that has since paralyzed your fancy. Recall that we are launched together, willy-nilly, on this extraordinary adventure of discovery. What kind of a universe is this? What are the laws of form?

I am going to tell you about a visit I made in A.D. 2000, in the company of some of your grandchildren, to one of the six regional universities at New York, Moscow, Tokyo, Delhi, Paris, London. There was rightly no world university since diversity is the life-blood of research and learning. This was of course a visit in the dreaming imagination, but imagination is the womb of reality.

---

* Such *relative* prophecies are often more reliable than absolute ones, in which the time required is frequently underestimated by a factor of from 3 to 5.

We attended a refresher course designed to provide an intensive massage and freshening of the brain, in the interest not of some totalitarian power remaking the human mind, but of the pure delights of the intellectual imagination. This was achieved by providing graded doses of excitement, by giving glimpses of the great vistas across the world of natural experience discovered during the second half of the century as the result of a new and more powerful conception of form. The course was called "The World of Form." It took a month, of which the last week was kept clear for discussion and debate. The final day of each course was devoted entirely to criticisms of the course by visiting Professors of Form from the other continental universities, and only those participants who submitted constructive suggestions for the improvement of the course were granted a diploma. The main feature of the first three weeks was a carefully designed series of some twenty stereoscopic films, with color and sound, supported by a spoken commentary.

When I returned from this visit I realized it would be almost impossible to convey to readers who, like myself, have grown up in the first half of this century, the science and the art, that is the wisdom, that had gone into the making of these films. It was as though the furthest insights of deep psychology had been put at the service of the most reverent love of nature and respect for the young mind.

All I can do is to give a list of the films that I remember and to describe one or two which made the most vivid impression on me. How I wish they were there for you to see now!

A. *Papilios*

B. *General Survey of Forms* (nebulae to atomic nucleus; the range of variation in time scale and in spatial scale; static forms and changing forms)

C. *Structure* (nuclear, atomic, molecular and crystalline; genes, chromosomes, enzymes, viruses; organic tissues and organs; the fundamental organic processes: multiplication, modification, growth, differentiation, etc.)

D. *History*

1. The evolution of species
2. History of the individual: embryo
3. History of the individual: person
4. History of the sense of form: in the species
5. History of the sense of form: in the individual

E. *The Emergence of Novelty*

1. The synthesis of nuclei and molecules
2. The emergence of life
3. The story of a mutation
4. The working of the human brain

F. *Forms of Behavior*

1. The patterns of normal, neurotic, and psychotic activity
2. Adolescence, love, mating, reproduction
3. The unity and variety of human behavior
4. An historical episode from *War and Peace*.

This program may have as much meaning for you as a list of Beethoven symphonies and quartets would have had for Beethoven's grandparents. I can only do my best to make some of them come to life.

The aim of the first film was to soften us up for what was to come by startling us with unexpected beauty. I shall never forget it; we spent an hour hunting tropical butterflies, ending with the chase and final capture of a fine specimen of *Papilios,* the most glorious of all. I cannot show you its colors, but I can let you know how it affected Alfred Russel Wallace * when he first saw *Papilios:*

> I had begun to despair of ever getting a specimen, as it seemed so rare and wild; till one day . . . I found a beautiful shrub . . . and saw one of these insects veering over it. . . . The beauty and brilliance of this insect are indescribable, and none but a naturalist can understand the intense excitement I experienced when I at length captured it. On taking it out of my net and opening the glorious wings, my heart began to beat violently, the blood rushed to my head, and I felt much more like fainting than I had done when in apprehension of immediate death. I had a headache the rest of the day, so great was the excitement produced by what will appear to most people a very inadequate cause.

This film left us glowing with the naturalist's love of nature, his sense of the beauty of life in its natural setting, and of the challenge to man who so seldom manages to live beautifully.

I need not say much about the next. It gave an orthodox but fascinating glimpse of most kinds of spatial form, from nebulae and solar system; earth, mountain, river, and lake; to the organic realm and man. It was designed to insinuate into the unconscious mind of the audience many parallels and principles that were made explicit in later films. Nor need I say

* Who formulated the principle of evolution by natural selection simultaneously with Charles Darwin. See Wallace, *The Malay Archipelago* (New York: Macmillan, 1898).

anything about the film on Structure, we will come to that later.

There followed a remarkable reconstruction of the entire evolution of life from the earliest single cells to Homo sapiens. I heard that they found it difficult to get an artist-zoologist to do this one, partly because of the persisting disapproval of "speculation." But it is obvious that a concrete guess regarding the noises made by early mammals, or the distribution of the hair on prehistoric woman, or any other such problem, is of immense value in provoking critical discussion. Indeed, it was only when Blackrow was preparing the first script of this film that he realized that all the evolutionary biologists before him had forgotten to take the famous "Blackrow factor" into account and so gained his Nobel Prize in 1980. Of course we do not yet know what the Blackrow factor is, but it radically changed the interpretation of evolution, adjusting the Darwinian thesis and entirely changing its significance.

D(2) and D(3) were really one continuous story, the history of a single human person from fertilization to death. I hardly dare to hint at the quality of this double film to readers that may not possess the objectivity and acceptance of facts which are supposed to mark the "truly philosophic or scientific mind." For the first showed in some detail nothing less than the coming into existence of a man: ECCE HOMO. The wandering of the seed, fertilization, implantation, cell division, the growth and differentiation of the embryo, the first functional activities and movements, delivery, the first cry. Like Russel Wallace seeing *Papilios* I could hardly breathe, the intricate drama so gripped me; here was the majestically harmonious unfolding of a necessity that had followed its path a myriad times in a myriad wombs, unwatched, ununderstood, uncontrolled by anything other than the laws of form. As the film continued and one saw hierarchies of patterns within patterns, of cells, tissues, and organs unfolding simultaneously, I felt the strangeness of it all: this developing form might be fated to become a Buddha or a criminal, a Cleopatra or a monster, a Newton or a cretin.

But now we saw the hideous little prehuman face, growing more terrible as it became more definitely human through the fifth, sixth, and seventh weeks, till I could hardly keep my eyes on the screen. While the mother is all loving preparation, this abominable little creature is beginning to stretch and exercise himself. Here was all this going on and the mother was sublimely innocent of any intellectual understanding of what was happening, as innocent and ignorant as were the first couplings of male and female when sexual union began some hundreds of millions of years ago. Perhaps we were, even in A.D. 2000, really still just as unaware.

Up to now the film had stressed the unconscious aspects of the natural process of growth and differentiation. But now there entered that strange factor which is the almost intolerable burden of our species, the one which we are most of us nearly all the time seeking to escape: the awareness of a responsibility for our own actions and for existing in a world where there is suffering, where living things must endure painful frustration. This has much to do with spatial form, for watch with me, with your mind's eye:

We see the face and naked body of one person, a man, developing from the first cry to the last. At first we follow happily, with sympathy and hope, through the first weeks, months, and years. All is growth, promise, unfolding charm, potentialities beginning to be realized. The face is harmonious, the body healthy and well proportioned. We approach adolescence, and now rather suddenly—what in the name of human justice has happened? The face is still unlined, but it shows a deepening distortion, a vein of fear and cruelty—or is it shame and guilt?—of which there was no sign a few moments before. Soon he is forty, fifty, sixty, lined with care and hate, and the body is bent and twisted. Hands and feet that were noble are now knotted and stiff. A human life, once full of promise, is over, and it ended in disharmony.

I was told that this film was actually constructed from photographs taken every few days throughout the lifetime of a man who was born about 1920. It showed the true story of a human life, mainly in changing facial expression, in the course of half an hour.

We had seen the fate of one human being, and it might be any of us. For did this man make himself? At what moment in that half hour did moral responsibility enter? At 11:10 when he was at school and betrayed a friend and "felt he had done wrong"? Or at 11:14 when his first wholenatured love faded, and he began to seek his own pleasure instead? Was it his parents who made him weak and cruel? At what point in his life did he acquire that "full knowledge of the consequences of his actions," which is a legal criterion of sanity? This film left me choked with questions that no Professor of Religion, of Morals, or Philosophy, or Science who has ever lived could begin to answer. Nor Plato, nor Freud, nor any other sage.

Some of these remain with me: If heredity and environment together in some degree fix the conditions of every human life, at what ages and in what sequence does the person become aware of those conditions? For without awareness there can be no responsibility. Is there a necessary sequence in the development of awareness which all normal persons pass through, until they stop somewhere? If so what makes a saint or a Goethe go on continually deepening and enriching his awareness until his last day? If there is any meaning in "should," "shouldn't" one develop understanding as well as moral character? If so, why have the great religions betrayed that responsibility?

But we must return to the films. The same afternoon the story of a woman's life was told in the same manner. This time it was a well-known figure of the nineteenth century, and we were asked to guess who it was. We saw the infant, the child, the young girl, and then toward sweet seventeen we saw that she was rather a tomboy and not quite as sweet as one expected. Those lines that promise the future were slow in unfolding; perhaps there was a lack of balance in develop-ment; but she was a fine person, certainly very much of a woman and quite a bit of a man as well; intensely vital and intelligent; living fully, a personality, indeed a female Don Juan, yet productive both as parent and as writer. As she grew old, as grandmother, there was in her face all the compassion and understanding of a woman who has lived richly and grown ripe. She was at the end as wise and serene as the old Goethe, as her letters prove. A disharmonious childhood and youth had been followed in maturity by a rare beauty of spirit. One could not help loving her. She redeemed the morning. The form of a human life could display spiritual growth. It was, of course, George Sand.

D(4) and D(5) took us away from these intimate realms. The first traced the changing awareness of form throughout the history of all the civilizations, showing how superficial cycles of youth, maturity, and degeneration (overemphasized by Plato, Henry Adams, Spengler, and Toynbee) are set within a more profound secular development. The sense of form of the different civilizations was revealed; we looked out on the world with the eyes of an Egyptian, a Cretan, a Greek, a Roman, and so on, ending with the vision of the structural scientist of A.D. 2000. I particularly enjoyed watching the sequence of architectural styles, from the prehistoric and antique through the centuries to the contemporary modernistic and finally the structural school of the late twentieth century.

This was complemented by the story of the development of the visual sense and space-image in the contemporary human infant. We saw the early lack of co-ordination between the different modes of sense perception and between these and the motor responses. The baby has, so to speak, half a dozen different effective spaces: a mouth space, a touch space of the hands, a hearing space, and so on, and these are only gradually assimilated to one another to form what the adult regards as the "real space" in which we think that we live. Thus from one point of view the space of length, breadth, and height in which you are holding and reading this book is a complex image fused out of many primitive and partial images, and the value of this single image of space is that it allows us to organize our responses in a more effective manner.

The film showed how fishes, insects, birds, and dogs form very different working images of their situation, the electric fish, for example, sending out pulses of electric current, and the bat pulses of sound, which are distorted or reflected by objects around them. So we learned that, from this point of view, spatial form must be regarded not merely as an objective reality,* but also as a schema in our brain-minds, the result of fifty thousand years of social development of a species for which the sense of sight is of special importance, and of five years of personal development in our own childhood. And this schema is still developing. It is very different today from three thousand years ago, and may at any time undergo further transformations. For example, we may come to think not of discrete atoms set in a continuous space, but of space itself as being structural, in fact as being nothing more than the relations between particles.

Indeed this film showed how quickly the image of space changes for the scientist of the Western world. From 1920 to 1950, approximately, he had a strange idea of a kind of timeless space-time, an eternal block of four dimensions, essentially continuous and symmetrical. This proved to be a mistake partly due to people taking too literally some epigrams by Einstein and Minkowski which were only valid in a restricted sense (Einstein later repudiated these misinterpretations). But that did not prevent the image of a continuous symmetrical space-time being in fashion for several decades.

Then in the 1950's what had been clear to less specialized thinkers suddenly dawned on the space and time experts, and their image of "space" became within a matter of years structural and subject to one-way change in the course of time.* *

* "Space" retains some properties of "objective reality" even if it is also an image in our minds.
* * This is the history of space-time ideas as written in A.D. 2000. I shall be interested to see if it is correct.

Space became once again the domain of three dimensions in which irreversible and historical processes occur. It was instructive seeing how the dominant views of the specialists in branches of exact science change every few decades like those of the doctors!

We have nearly reached the end of this description of the course, but I must mention one or two more. I was delighted with the group on the Emergence of Novelty, how new forms come into existence and how this involves the interplay of law and chance. It was made clear that the overstatistical mood of the period 1930–50 had led some of the greatest physicists to suggest that "everything is statistical." That is as bad as saying anything is due entirely to heredity, or to environment, since both are always present. There is law and order everywhere *as well as* varying degrees of disorder.

I could not entirely follow the film on the emergence of life. But it was clear that there are various stages in this process, that in A.D. 2000 it was much less "miraculous" than it had seemed to be in 1950 or 1900 because the laws of form were better understood, and that the problem was really similar to others which had previously seemed quite different, like the synthesis of protein, the working of the brain, and so on. I remember that we saw the highly magnified chemical components of a droplet of slime which happened to lie in an electric field and near some magnetized iron—we saw these chemical structures being polarized and twisted and then massaged by quanta of sunlight until they began first to pulsate, and then to spread their pulsating structural pattern across the droplet, till the whole was throbbing rhythmically. This was fascinating: we were shown life as a condition of pulsation which tends to extend itself, and we saw this happening.

If I remember right this droplet was composed of a complex semifluid material which turned into something like protein in the process of beginning to pulsate. Here I am using the term "pulsation" for a cyclic or rhythmic process which none the less produces a progressive result, like the cyclic beating of the heart pumping the blood one way. The synthesis of protein involved pulsation, as indeed do nearly all organic processes. This was one of the crucial steps toward the living cell, the first rough blueprint of life, as we shall see in another chapter.

That was one kind of novelty. Another, equally dramatic and much nearer home, was the working of the human brain as it formed a new ordering of the plastic records of experience. I shall also have more to say of this later. But the subject of the film was the mode of working of the cerebral cortex during the development of a new idea, using "idea" in the broadest sense from any hunch, say how to fix a rattling window, to major creative efforts like the generation of a symphony in the mind of a composer. Here again we learned how it was the discovery of the laws of form which overcame the difficulties that had so puzzled the brain physiologists of the first half-century.

Now for the last film. At first I imagined that it was an attempt to show the dangerous quality of mass activities when men (particularly men!) get together and let one emotion or one idea or one person dominate their actions. The film was a study in the spatial patterns of social behavior: how a large number of individuals may come together, accept one over-riding principle, and do strange things in unison. I was struck by the fact that a passage which Tolstoy wrote in 1869 had been considered suitable for an educational film in A.D. 2000.

It started with a series of apparently disconnected shots of families in Western Europe at the opening of the nineteenth century. In each home there is a young lad shown in the setting of domestic life and the ordinary activities of the period. Then suddenly all these men start collecting in smaller and larger groups, putting on uniforms, learning to use rifles, and fitting themselves into a hierarchical discipline. It is 1812, some invisible person gives a sign, and they all, many hundreds of thousands of them, start marching eastward. It becomes very cold, a large proportion fall by the roadside, and eventually another signal is given and those that are left start marching westward. Nothing has been achieved and countless lives have been thrown away.

One was left wondering, with Tolstoy: "What does all this mean? Why did it happen? What were the causes of these events?" Then I realized that the film might have a special role in the course. It was perhaps suggesting that the unique events of human history lay outside the scope of ordinary scientific methods. The influence of a Napoleon, a Churchill, or a Stalin, could not be measured because there was no standard of comparison, no control group, and there could never be. Unique events lay outside "science," in the strict sense of the word.

Now that I am back in 1953 I ask myself: Was this dream nonsense or significant prophecy? No one can know today.

In 1953 there are already in existence films showing the savage fighting of a breed of mice which are frightened of the light; the latest researches in astronomy; animal, bird, and insect life; life under water; the courtship of the praying mantis; volcanic eruptions; the world of symbolism and dream; and so on. It may require two generations to pass from these scraps to a general survey of this universe and of human existence within it in terms of the idea of form. It will probably take fifty years of research to explore those vistas that I now dimly imagine.

IV.

Atoms, Wholes, or a New Look?

The conflict of atomism and holism is overcome in the idea of structure

EACH period in history is marked not merely by its special achievements but also by some characteristic neglect. A school of thought which focuses attention on one aspect of experience may be almost unconscious of some complementary aspect, and what has had a high light on it in one period may be forgotten in another. I am suggesting that an important blind spot of the present time is the failure to recognize the significance of form as a key to the understanding of natural process. In the ancient and medieval worlds form—in a vague sense—was recognized by many thinkers as being of the highest significance. My argument is that it will be again tomorrow, but this time with a clearer meaning. The Gestalt movement in psychology is an anticipation of a more precise and comprehensive method which has still to be found.

Here is an illustration of what I mean. Since about 1925 the word "pattern" has become fashionable in many branches of science. Yet one might almost say that the word has come in because the idea is still missing. No one knows exactly what is meant by "pattern"! At least not an adequate idea, because no laws of pattern have yet been established. I can imagine the historian of science saying to his pupils in the years to come: "It is really curious. The 1950's were rather proud of their standards of mathematical and logical precision, yet we can find no record of a logician or scientist trying to clarify the concept 'pattern' and to formulate its possible laws. I propose therefore to set as homework for this vacation: What the scientist of the 1950's thought he meant when he used the word: pattern."

Of course the fact that one can already think like this may indicate that scientific thought is nearly ready to achieve a deeper understanding of forms, structures, and patterns. One might think that a new insight of this kind would mature in ten years, but fifty would probably be nearer the truth. It is not merely that inertia, indifference, and prejudice always weigh heavily against new ideas. Max Planck wrote: "A new scientific truth does not triumph by convincing its opponents and making them see the light, but rather because its opponents eventually die, and a new generation grows up that is familiar with it." More than that, the new ideas themselves grow very slowly and can only mature when their time has come. But perhaps the time is nearly ripe for a new conception of form.

Nearly, but not quite. Some years ago I met a distinguished biologist, a man with a high reputation for original thought. In his presence someone used the phrase "the problem of form," and I heard him mutter sarcastically: "What is the problem of form, anyway?" I thanked him innerly; he had shown me two things. Firstly, that the time was not yet ripe, for he was one of the most enlightened; and second, that one must express the problem so clearly that it could not be evaded.

Sharp criticism is the life blood of science. When one could formulate the problem satisfactorily so that his attitude would be impossible one would have got somewhere.

I think I have made some progress toward formulating the problem in mathematical terms, which I will here try to translate into ordinary language.

The physics of this century has used atomic ideas with immense success. These ideas lack any explicit principle of form, or shape, or regularity, the ultimate particles being conceived as independent centers of influence, each capable of existing separately and possessing its own properties, such as mass, electric charge, etc. THE PROBLEM OF FORM is then:

In an atomistic universe how can regular forms develop? Would they not be at best highly improbable?

Or to invert the question:

If this is really a universe of formative process, and regular forms hold the clue to understanding, how is it that the doctrine and mathematics of atomism have been so remarkably successful?

This is one of the most penetrating questions of natural philosophy, but I believe the next generation must answer it. The issue is: when are scientists to think *analytically* in terms of the smallest parts they can find, and when *formally* (in the very old and new sense) in terms of the changing forms and patterns that they actually observe? For they never see or photograph a single particle at one spot, but only extended spatial patterns. The answer given by exact science to this question will exercise a wide influence on thought in all realms.

Some may be inclined to say at this point: "The whole contrast of atomism and form is spurious. It is agreed that the atoms make up the pattern, and the patterns are composed of atoms. There is no genuine problem here." But we shall see that this attitude is wrong. There is a sharp contrast between the two ways of thinking as they are used now, and if we want to attain deeper understanding we have to discover how to reduce that contrast and to find one way of thinking which can do justice both to the atomic aspect and to the form aspect of things.

The success of particle physics since 1900 proves that the atomic aspect is unquestionable. But open your eyes and look out on the world. You may see a kitten, a flower, a tree, a crystal, a fiber, a cloud, the crescent moon, a painting, a person, or this printed page; the form aspect is there as surely as anything in human experience. Regular forms are not infrequent or improbable, they dominate nature. The problem is how to interpret or combine these two aspects so that they contribute to a deeper comprehension.

This is a difficult task, for two reasons which are really one. Firstly, since the time of the ancient Greeks, thinkers have shown a tendency to fall into one of two camps, which for convenience we will call the *Atomistic School* and the *Holistic School*. So the habit of separating the two aspects has become a vice that is hard to overcome.

These two schools have spent much energy challenging one another, and the heat generated shows that they must express two contrasted types of human temperament each of which cannot help disliking the other, because the other school emphasizes what it lacks, or has inhibited, or has refused to recognize. Since real achievements sooner or later speak for themselves, the heat of this battle proves that unconscious prejudices are at work.

This is perhaps the deeper difficulty; there seems to be an

inherent lack of balance in human nature which makes the
mind tend to incline either to one side or to the other and to
display a prejudice, normally unconscious but sometimes
becoming very conscious, either against atoms and particles,
or against wholes, forms, and unities of any kind. For *no
Western thinker has yet displayed a balanced attitude*. No one
has found an elegant way of keeping part and whole in
balance. The influence of every great mind has fallen on one
side or the other, and in doing so has somewhere distorted the
truth. This is an important fact, responsible for much intellectual frustration.

The Atomistic School may be represented by Leucippus,
Democritus, Gassendi, Newton, Boyle, Dalton, Rutherford,
and contemporary atomic physicists. It was Rutherford's
delight to chase the particles that Indian and Greek thinkers
had first imagined two and a half thousand years earlier, and
now every atomic physicist is doing it.

The Holistic School has had less fun and much less success.
In spite of the high reputation of their leaders: Aristotle,
Goethe, Bergson, the Gestalt psychologists, Whitehead, and
Smuts, they often appeared to be defending a losing cause.
Wise men, but nature seems to have given a secret tip to her
favorites on the other side: Newton, Dalton, Thomson,
Rutherford, Millikan, and all the others in our own day.
Beside their achievements Goethe and Whitehead often had
the air of merely expressing pious hopes. But the question is:
Are the atomists the permanent favorites of nature, or will
their luck desert them?

The atomistic view and a philosophy of life built on it are
delightfully described in Lucretius' *De Rerum Natura*. A
recent survey of the opposed outlook is given in Smuts' *Holism*

and Evolution, which is mainly valuable in showing how a nonspecialist mind can point to the blind spot of his own time. This book is neither precise nor clear, but it is probably still too early for a mature statement of the theme. Goethe would have valued the book, and the scientific philosophers of tomorrow will recognize it as the anticipation of a crucial scientific problem. But exact scientists today find it vague and unsatisfactory.

Let us examine the two doctrines more closely. At first sight the one is as clear and sharp and fertile as the other is vague and dogmatic.

The classical atomistic doctrine asserts that the universe is made up of ultimate particles, each of which is simple, indivisible, and permanent. All observable changes are due to the reversible spatial rearrangements of these particles resulting from their motions and mutual influences. The particles although small must be of a finite size or effectively occupy a definite volume of space, since a finite number of them make up ordinary objects. Moreover if atomism is to work there must be very few different kinds of ultimate particle, for the aim is to simplify our view of nature. The complexity of observed phenomena is to be accounted for as the result of the motions of units which are each simple and permanent.

This program had the great advantage that it gave physical measurement and mathematical reasoning something to grip. Nature becomes a kind of geometry representing the motions of the points at which the particles are situated, and temporal change is reduced to movement. Democritus, Newton, and Rutherford all thought and worked in terms of this powerful conception of the nature of change. Yet in some way which is not yet quite clear it is inadequate. Perhaps it is too weak, and does not sufficiently restrict the range of permissible movements.

Beside the achievements of atomism the holistic view is an undeveloped intuition, a mere hope. It regards the universe as a great hierarchy of unities, each following its own path of historical development. Each pattern, whether it is a crystal, an organism, a community, the solar system, or a spiral nebula, possesses its own internal order and is part of a more extensive order, so that the universe is recognized as a System of systems, a Grand Pattern of patterns. Every whole with all its parts is subject to developmental changes which cannot be adequately represented as the mere reversible motions of independent particles.

The favorite model of the atomist was a chaos of moving particles, like the molecules of a thin gas at a high temperature flying in all directions and subject only to occasional random collisions. He therefore preferred to neglect as far as he could the anomalous facts that do not fit this model, the islands of order such as crystals and organisms. It is curious how obsessed many of the mechanical atomists around 1880–1900 were with the restricted corners of physics which were their immediate concern. A mathematician would say that they did not realize that they were dealing only with "highly special limiting cases of a more general expression." For it seems that order is quite as prevalent as disorder in the universe of human experience.

Meantime the holistic thinker's model of the universe was an organism, in which every part is harmoniously related to the processes characterizing the system as a whole. And he in his turn was either blind to or became cross about those unfortunate blemishes where the divine hand seemed to have slipped, and let in some feature that justified the atomistic heresy. (See Goethe on Newton.) While the atomist saw the finger of God in the simple geometry of atomic and planetary motions, the pattern thinker held that to be a betrayal of his own true God as revealed in the organism of the universe. The one group found perfection in analysis, atomism, precision, and quantity; the other in form, order, and unity; and they seldom appreciated the qualities admired by the opposing group.

This may seem to be an abstract matter of little practical concern. But let me put this question: In your ordinary daily life or professional activities do you tend to think intuitively, using a direct sense of general situations and the relationships involved, or analytically, by breaking everything down to its components and trying to start with the detailed facts? If you have not a bias one way or the other you are a very rare kind of person. One may say, generally, that the "feminine," artistic, poetic, inventive, and imaginative component of human personality uses the first or holistic approach, while the "masculine," analytical, classifying component uses the second, or atomistic method. But we all to some extent use both; it is a matter of emphasis and degree.

And yet every one of us is stronger on one side or the other, and is therefore subject to an unconscious bias and resentment. Philosophers and scientists have constructed great systems out of their preferences, and in doing so have cast veils over the truth.

From 400 B.C. to around A.D. 1925 these two schools competed for control of the Western mind, and through the centuries the dominant mood of the educated classes and of the universities has repeatedly swung from one side to the other. A study has been made of all known cultures showing a continual oscillation between atomistic and holistic attitudes. But it is a mistake to press this kind of analysis too far. The two methods are complementary, they are in perpetual interaction, and their significance changes as science advances. Indeed I believe the time has come when the traditional form of this conflict must disappear. Since about 1930 science has entered realms where this contrast must be overcome.

The evidence for this suggestion lies in a dramatic situation which has recently developed in the interpretation of fundamental physical experiments.

Between 1870 and 1900 two of the greatest philosophers of exact science, Ernst Mach and Henri Poincaré, considered that atomic theories should be regarded merely as ways of inventing mathematical equations to represent observed quantities, and that it was wrong to think of the atoms as in any way "objectively real." But by 1920–25 the paths of individual atoms and particles had been made visible to the naked eye, and every experimental physicist believed in the real existence of the electrons and protons of which atoms are composed. The prevalent view was that Mach and Poincaré had been proved wrong, and Democritus finally justified.

What a platitude that pride comes before a fall, but how startling it is to see it happen punctually, when assurance has become absolute! Perhaps there is a divine agency, AJOTE, Aesthetic Justice on the Earth, looking after the timing of these matters. Or maybe it is that while one method is being exploited to the limit, the complementary method must be neglected—until it cannot be any longer and comes back with a rush.

That is what began to happen between 1925 and 1932, just when physical atomism was thought by many to have been finally established beyond question.

The first sign of difficulty was given by the wave theory of de Broglie and Schrödinger in 1924–5. For the particles now turned out to be quite unlike the little Democritan atoms, they stretched across space like an Atlantic wave pattern reaching from the East Coast of America to the West Coast of Europe. Notice the crucial point: the particle has become an extended pattern, though what that means is not very clear, if it is still also a particle.

The second signal was provided by the progressive discovery, starting in 1932 and still continuing today, of a complex series of ten or more kinds of *unstable* particles (some of them called *mesons*), which appear and disappear during the collisions and interactions of the more permanent ones. The Democritan atoms were simple and spatially minute, but persisted forever; the post-1925 particles are mostly complicated and unstable, but extend spatially as far as the boundaries of the experiment. The atomic doctrine had been transformed with a vengeance, and by unchallengeable experiments. I can hear Mach and Poincaré in the Elysian Fields saying, "It looks as though we might be partly right after all! The atoms seem to be merely the markers of patterns."

In another realm the traditional conflict of atomistic and holistic thought had already begun to fade out somewhat earlier. The mechanism-vitalism dispute had been at its height around 1910–20, and just at that time a new less polemical and more balanced attitude began to develop. This sought to escape the old subjective prejudices and tried to identify the objectively given structural pattern in everything, whether atom, living cell, organism, or mathematical system undergoing some transformation.

For convenience we will call this New Look at nature the

Structural School, though it is not yet widely recognized. It is risky to link individuals with so recent a trend in thought, but it seems that Bertrand Russell and his followers in logic; many workers since the middle of the last century on the foundations of mathematics and on molecular and crystal structure; organicists such as Bertalanffy and Woodger in biological theory; and various recent schools in psychology and sociology all share one aim: to identify the effective pattern of relationships in some natural phenomenon or mathematical system.

The strength of this attitude lies in its objectivity and lack of prejudice, based on the desire to eliminate whatever is irrelevant in inherited ideas and to retain only what is necessary and sufficient to cover the structure of the facts.

The weakness of this structural school of thought is the other side of its objectivity: an exaggerated valuation of skepticism, the absence of any unquestioned conviction about nature, and the lack of constructive ideas to guide experiment and theory. For no major advance in science can rest merely on skepticism, caution, and observation. A conviction of the unity of nature expressed in the form of a particular idea is necessary, for the intimidating complexity of the facts can only be overcome by a person who believes that he holds a clue. Without some conviction or provisional belief regarding the specific form of the simplicity of nature, that is without some guiding idea, no experiment can be designed and no calculation undertaken. The Structural School is in danger of thinking that the elimination of error and prejudice and the collection of facts are the sole objects of scientific endeavor. But we can only know that we have got rid of an error when we have been able to advance to a simpler and more comprehensive truth, and the idea must come before its justification.

An important element is therefore lacking in the structural doctrine. This missing feature is easy to describe, and difficult to justify.

The task of science is not merely to identify the changing structural pattern in everything, but *to see it as simple*. Science starts with an assumption which is always present, though it may be unconscious, may be forgotten, and may sometimes even be denied: *There exists a simple order in nature; a simple way of representing experience is possible; the task of science is to discover it.* The true aim of science is to discover a simple theory which is necessary and sufficient to cover the facts, when they have been purified of traditional prejudices.

When this scientific affirmation is forgotten or denied, science withers in a desert of specialisms. Listen to some voices from this desert:

"What do you mean by simple? The latest and most powerful physical theories are far from simple." Science is in a bad way when it is necessary to try to explain to a skeptic the meaning of "simple." For all intellectual processes depend on the operation of the aesthetic sense which recognizes an elegant ordering when one is presented to it. This sense is prior to reason and cannot be justified by analysis or interpreted by definitions.* The latest and most powerful physical theory (quantum mechanics) appears simple to a certain highly mathematical kind of mind, but its baroque elegance is a smoke screen which conceals some rather shabby patches.

* No definition of simplicity given today is likely to stand tomorrow. The complexity of a theory may sometimes be measured by the number of independent postulates and of independent parameters which it employs. But this is seldom easy to discover.

Once a theory with classical simplicity and elegance has appeared, the claim that quantum mechanics satisfies the sense for simplicity will be forgotten.

Here is another voice: "The facts are as they are and can never be made less complex." I do not remember who it was that displayed this innocent failure to understand what the Western mind has learned by hard effort during the last hundred years: that everything that we at first naïvely regard as "the given facts" are actually interpretations biased by the organic situation of the human brain, by traditional conceptions, and by the recent experiences of the individual. Science does not begin with facts; one of its tasks is to uncover the facts by removing misconceptions.

And yet another: "But physical nature is very complex, for we have discovered some twenty different kinds of particles. The physicists of 1920 knew only two or three of these." To which the active scientist, whose will that discovery shall continue is greater than his skepticism, answers: "This apparent complexity is partly due to our current way of thinking, which is clearly inappropriate. We have not really *discovered new particles;* we have discovered new facts which prove the inadequacy of our old ideas of particles. Something is wrong in our present conception of atomism; we must discover how to improve it so as to eliminate the need for calling every discovery a new 'particle.' A novel principle is needed which will restore simplicity to our physical picture."

The situation resulting from the 1930–50 challenge to atomism is so dramatic that one can enjoy it year by year without taking sides. The next ten years can scarcely fail to bring something fresh that will send one side or the other staggering. It would be most surprising of all if there were no great surprises soon. For nothing really new has happened to fundamental physical theory for twenty-five years, and that is a very long time for physics in this century of rapid discovery.

But one can enjoy it even more by forming a judgment and taking sides. It does not matter if one is right or wrong, for the interest lies in learning whether one has judged rightly or wrongly, and why.

Every reader of these pages can participate in the drama. Science is not a tedious affair of examinations, degrees, salaries, fears and jealousies, personal gossip and intrigue, honors and dishonors. It is the commando of the incomparable adventure of discovery, and you, the reader, can watch a raid into the unknown. You can make your own guess, with me or against, regarding the shape of the unknown, which I believe will become known soon.

This is an unorthodox procedure, not to be imitated except after due preparation. But it is fun, and we can enjoy scholarship and research, forming judgments, taking sides, and being good losers.

Here is the bet: *I offer odds on that the present embarrassment of atomism will by 1975 be recognized as due to the neglect of form.* That the complexity of current particle physics will be overcome by a mathematical method which, in a manner that today cannot be foreseen, combines some aspect of the old atomism and some property of form into a new single doctrine. And that physics will then use only one kind of permanent particle.

Now in a world where the true principles of education were understood, one would be able without loss of face to add the following suggestion: "If you appreciate the point of this, go to your friend or your teacher who is an exact scientist and

see if he will bet against you. Choose your side, select the odds
you will offer, write it down in a carefully prepared formulation, and every year check up on it. And do not let any
pompous person tell you that the advance of knowledge is a
grave matter only for specialists. The outsider often sees more
of the game, no one knows where the next idea will come from,
and science is above all an adventure. Even a nontechnical
essay like this may contain a clue, and help to push on the
advance of science."

These two ways of thinking have been at loggerheads for
over two thousand years, and prestige has oscillated continually
from one to the other. My suggestion is that this age-old
controversy has now reached a final crisis in the decisive sphere
of fundamental physics, and that nature will at last settle the
case by demonstrating that *neither* technique alone is adequate
for penetrating her secrets. Newton and Dalton will be
deserted, but the holists will be also, for nature's new favorites
will be the School of Elegant Structure, combining an intuition
of the whole with the analytical recognition of detail.

But the crucial question is: What new way of thinking can
retain what is right and drop what is mistaken in both schools,
and by doing so achieve a greater clarity?

Let us take the atomists first. There is no doubt that unique
physical points, or ultimate particles, are necessary to define
the spatial patterns observed by the physicist. In this sense
nature is granular, and physical systems are composed of
particles. So far so good.

But the atomist, not content with this, goes further and
assumes that the ultimate laws for which he is looking must
describe the properties of single particles. He tries to ascribe
the changes in the patterns to the physical properties of the

ultimate particles, their electric charges, masses, and so on. This assumption is unnecessary and may be a mistake. The laws may represent the complex changing patterns directly, not indirectly through "properties" ascribed to single particles. In the ultimate analysis the particles may possess no properties! The only role of the particles may be to mark and anchor the patterns. On this view the particles are merely the keystones or focal points of the patterns.

Now take the holists. When they say "the whole is more than the sum of its parts" they are trying to express a valid principle, but they do it in such a manner as to make it obscure. What they should say is "The *laws* express the changes in whole patterns and cannot be expressed in terms of properties of single parts."

The holists are right in thinking that complex systems are important, for the laws describe how such systems change in course of time. And the atomists are right that discrete structure is important, for that alone distinguishes one system from another. But the holists neglect structure, and the atomists the properties of systems.

The New Look at Elegant Structure seeks to overcome the old conflict for it sees universal laws of formation and transformation, illustrated in varied systems each with its specific structure. The unity of nature lies in the operation of one general law of change, while the differences of particular systems lies in their individual structures.

If these suggestions are correct physics must shortly shift its emphasis from single material particles possessing masses, charges, etc., to the changing shapes of complex structures. Indeed, this is already happening. Schrödinger has suggested that the philosopher of the past would say that the modern atom consists of no stuff at all, but is all shape. But physics has not yet taken full advantage of this implicit transformation; the revolution is still in suspense. The old king, Atomism with particle properties, has lost his authority; the new king, Structure with formal properties, has not yet been acclaimed.

The old regime spoke mainly of *particles, forces, reversible motions,* and *conservation laws;* the new must learn to speak of complex arrangements of different kinds, and their *symmetry* and *asymmetry, one-way changes,* and *rhythmic pulsations*. Indeed the signal that the change-over is complete will be the issuing of an edict that one may not speak of "a fundamental particle" since that by itself is meaningless, but only of patterns of particles.

Three master problems urgently await clarification under the new regime.

1. The development of a physical theory unifying all the new fundamental knowledge of this century (Relativity Theory, Quantum Theory, Nuclear Theory). This may provisionally be called the *Theory of the Fundamental Particles.*

2. The clarification of the relation of inorganic to organic processes, throwing light on the properties of organisms. This is the *Theory of the Pulsating Structure of Organisms.*

3. The description of the processes occurring in the human brain during its unconscious and conscious operations, in the *Theory of the Pulsating Structure of the Brain.*

The achievement of the first would provide a more extensive foundation for the whole of exact science, that of the second would transform the problem of life, and of the third would assist the establishment of a unitary language overcoming, where this is desirable, the dualism of the "material" and "mental" languages.

Every scientist is aware of the challenge of these three great problems, but they are normally regarded as independent of one another. The first is for the particle physicist; the second for the biochemist, physiologist, and biologist; the third for the neurophysiologist and psychologist. At least so it is generally believed, and this is correct as regards the experimental investigation of the three problems.

But none the less the three problems share one important feature. In each case the task is to understand the properties of a complex system, involving many component parts. It is this which makes all three so difficult.

Until recently astronomy and physics were mainly concerned either with very simple systems, such as a sun and one planet or two particles exerting forces on each other, or with systems that were so complex that they displayed simple statistical properties, such as a fluid with a temperature or pressure in every small region. But these three master problems are concerned with the interactions within systems which are too complex to be reduced to simple situations of the traditional kinds. A new method is necessary for all three, and any method capable of dealing with complex forms may simultaneously throw light on each of the problems. What is needed in each is a way of making complexity seem simple.

So here is another suggestion: only a new scientific doctrine of structure and form can suggest the crucial experiments which can lead to the solution of the three master problems of matter, life, and mind. Many of the special sciences today require a way of making complex systems appear simple, so that a single observation can reveal something about the system as a whole. When such a method appears, it may be possible to apply it at once to a system of particles, to the internal arrangement of an organism, and to the working of the brain.

It has always seemed to me doubtful that the image of God as a Very Brilliant Abstract Mathematician could be right. For He would only need to be expert in complicated and difficult branches of mathematics if He had failed to invent a truly elegant way of designing the universe. Such a thought is blasphemy to the best scientific, as to the religious mind. If a human mind can yearn for a high degree of elegance God must surely have been there before him and have already achieved everything that man can dream. So I prefer a God of Elegance with standards as high as can be imagined, so that He need not be a particularly good mathematician. For the physicist this means a theory based on one kind of fundamental particle, with no properties other than its permanence, arranged in patterns obeying relatively simple laws.

It is appropriate to use the image of divinity in explaining a scientific doctrine because the scientist's a-rational conviction of the simplicity of nature is closely akin to religious faith. Indeed I suggest that these two are both expressions of the vital tendency toward order. It is inherent in our organic make-up that we seek a simple order, and if such a simple ordering or a tendency toward it were not present in nature no organisms could exist. For every organism is a system of mysterious elegance challenging the intellect to discover its secret. So I have here used "God" as a symbol for an unknown principle of order or elegance.

Physics seeks to penetrate the music of the atomic spheres, biology the harmony of the organism, and neuropsychology the melody of thought, and though they do not yet know it these three sciences may be seeking to discover the same universal principle of elegance. This principle must be simple and must define the character of change in complex systems. There seems to be little choice. It must surely express a *natural tendency toward simplicity* pervading all realms. Nature moves toward simplicity, and therefore natural law working in man leads him to seek simplicity. The human mind and its ideas are adapted to discover the order of nature because the mind is a part of that order and shares its ordering tendency. The human mind seeks a simple order, because nature does also.

Hitherto scientific thought has regarded change as fundamentally reducible to the *relative motions of entities*. In the future it may be necessary to revise this interpretation, or to generalize it, and to regard change as *changes of form*. Beneath the apparently haphazard motions of particles may lie a formative tendency, or tendency toward simplicity of form, order, and regularity.

V.

The Face of the Moon

*The markings on the surface of the moon
prove the reality of its history*

BUT now let us come down to the facts as we know them.
What is known about this universe into which we are born,
and what insight is provided by the idea of form?

When each of us grows up he gradually becomes aware that
he is partaking in the strange adventure of human life which
no one understands. He is one of two and a half thousand
million persons alive in the 1950's of the Christian Era, on the
surface of a globe about eight thousand miles in diameter
spinning on its axis and moving round its orbit at twenty miles
a second. Moreover the human species is alone in this adventure; no other known species is actively trying to understand
the universe, as man is. This is all very strange, when you think
of it. Why is it so? Or if *why* is too much to ask, *how* did it
come to be so?

The physicist and the astronomer make fascinating guesses
about the past, but are still uncertain on all fundamental
matters. The heavens are well mapped and the geometrical
rhythms of the solar system have been accurately measured,

but the secret of their history is still hidden. One day it will be known as surely as is the law of gravitation, but at the moment the young science of cosmology has to rely on precarious assumptions.

Here is one which may not survive very long. Today the distances of the nebulae are measured in "light years," the number of years it takes light to travel from the nebulae to the earth. But no one has ever directly measured the time light takes to travel in one direction. Does light really "travel," like a motor car or a plane? Must we assume that light "starts" from a distant nebula countless years before it "reaches" us? Light may not possess a true velocity, and the present way of speaking of cosmic distances may be misleading.

Then here is another weakness in present cosmological speculations. No one has ever been out to one of the distant nebulae. All our knowledge rests on the use of physical instruments on this earth and on our knowledge of the laws which these instruments obey. But we have not yet got a satisfactory theory of the properties of the particles of which these instruments are composed, and a tiny adjustment in our basic physical ideas might entirely transform our picture of the cosmos. Moreover the recent discovery of radio stars suggests that so far we may know only a fraction of the stellar universe. Cosmology should be prepared for surprises.

But some things are already clear. The universe contains vast nearly empty spaces, and relatively small islands of matter. Moreover it displays a considerable degree of form, regularity, and orderliness. The telescope reveals neither a random chaos of particles nor a uniform continuum. For example, there are great galaxies of spiral structure, disk-shaped clouds, double stars, and one solar system which displays a high degree of internal order. This is Point One regarding human destiny: we awake to consciousness in a universe which is not chaotic. There is structure and form, about which we can think and speculate and calculate.

This primary fact is often obscured by mathematical abstractions and by loose use of words such as "random," "disorder" and so on. In a void universe, wholly without form or structure, there could be no intelligible process, no history, no life, no perception, and no thought. We need not try to think about it further, because nothing can be thought about such a universe.

But once a specific form or structure is perceived, how much there is to think about it, how much that one is inevitably led to think about it! A nebula with spiral arms evokes a deeplying tendency in our mental make-up: we think it must be rotating. A plane figure with an axis of symmetry perpendicular to it and structures, like the spiral arms, which discriminate one direction round the axis—such a figure compels us to think of rotation. Here a spatial form implies a process, or seems to. Moreover we are led to make a further inference: the rotation must have produced the spiral structure, the process which is suggested by the form must have produced the form.

Thus when we of the twentieth century observe any structure we are led to think first of all about its past rather than its future. We do *not* think: "This structure has existed from the beginning of time as it is now, but at some moment in the future it will be broken up into its constituent parts." That would lead nowhere. But we do think: "This structure was formed once. It is a record of past history, for its structure tells us something specific about a particular process in the past, the process which brought the structure into being." *

When the twentieth-century astronomer meditates on the cosmos he usually finds himself assuming that its present structure must have been the result of a process of evolution from some earlier more uniform or less structured condition. He tries to assume an "original" state of rather uniform but unstable chaos, and to see how its present form can have developed from that. It seems that he is driven to interpret any existing form as the record of a formative process in the past which brought the form into existence.

This contemporary tendency is the more striking because, at least in its present guise, it is not an innate characteristic of the human or the scientific mind. Isaac Newton, whose scientific methods in certain respects set a standard for all time, did not think historically when he was concerned with mathematical philosophy. The purpose of his "System of the Universe" was to reveal the timeless cycles, or unchanging geometrical transformations, of the solar system and the heavens. His mathematical genius was the more reliable because it accepted this strict limitation, and entirely neglected the historical problem presented by the specific form of the solar system. Newton explicitly recognized the challenge of the facts that the motions of the planets and of their satellites (with very few exceptions) are all nearly in one plane and in one direction, and that the sun and the planets all rotate in the same direction. But he held that the explanation of this was

* This historical and formative bias, which leads us to look for a past process of formation when we perceive any definite form, has a philosophical justification, which here can only be hinted at: asymmetries can disappear and symmetry can develop spontaneously, but existing symmetries cannot disappear without a cause.

a problem for religion, not mathematical philosophy. He is definite on this point:

> . . . it is not to be conceived that mere mechanical causes could give birth to so many regular motions. . . . This most beautiful system of the sun, planets, and comets, could only proceed from the counsel and dominion of an intelligent and powerful Being.

There could be no clearer demonstration of a transformation of the scientific mind. In 1686 the greatest natural philosopher of all time banished history from physics; in 1953 every scientist reads history into natural forms.

For Newton God was the one and only formative principle; everything else was the timeless geometry which God had set in motion. With marvelous humility and foresight he realized that his system might prove inadequate and serve as a guide toward "some truer method of philosophy." But he did not speculate on what that more comprehensive method might be, and the influence of his achievement is still so profound that fundamental physics has not yet entirely overcome his rejection of history and preference for cyclic and reversible processes.

Yet the sense of history is now deep in our mode of being and thinking. It is scarcely possible today to consider any problem without searching for a continuity of development leading from an earlier to a later state, and this often means from a relatively complex or formless to a simpler or better defined state.*

Thus many cosmologists assume that "at the beginning" the universe was relatively formless, say a vast volume of hot gas

---

* Mathematically: from a state requiring more independent numbers to describe it, to one requiring fewer.

and radiation, possibly of the simplest material of all: the nucleons (neutrons or protons) of which all atomic nuclei are composed. This is a reasonable working assumption to start with because it is the least arbitrary, and it takes a large burden off Newton's Being, converting it into a problem for exact science.

The philosophers and theoreticians of science have scarcely realized what an immense task has thereby been put on the laws of nature. The postulated initial condition is a vast sea of single particles spread through space in an unstable non-equilibrium state. The present condition is everything we have now: all the subtlety of human observation and experience, the myriad forms of organic life and of human culture. And the laws have to describe how the initial became the present condition. The daring of Prometheus was nothing to this, for here science has stolen the creative task of the gods. It is clear that if science is to succeed it must employ laws of tremendous power.

The normal scientific method is to cut off single problems and to cope with one at a time. Cosmology has done this in its own generous fashion with a curious result: it has cut off about four thousand million (4.10^9) years of the past history of the universe and calls that "the age of the universe." When cosmologists assert, as they have recently, that there is evidence that the universe is about four thousand million years old, one seems to be invited to think: "So that was when the universe was created." Quasi-religious statements of an absolute kind concerning the totality of existence have reappeared as a kind of boundary condition of science. A boundary stake has been fixed at a particular point in the vast dimension of time, and on it is written: "Thus far science; on the other side God."

But perhaps the stake now called "the origin of the universe" only marks the provisional exhaustion of the twentieth-century Prometheus, and another scientific generation may repudiate it. We have left Newton behind, but we may not have gone far enough.

Only a short time ago it was reasonable to imagine that theories of the evolution of the universe were so far from the check of direct observation as to be likely for many generations at least to remain highly speculative calculations about processes which could never be approached more directly.

This assumption may prove to be wrong, and for a reason which is delightfully concrete and unexpected. We may have a beautifully clear record of conditions rather near that zero time-stake at our doorstep most evenings *in the face of the moon. For it is probable that many of the structures visible on the surface of the moon were formed at the same time as the earth and the solar system (around three thousand million years ago) and have changed very little since.* The face of the silver moon may be a mirror reflecting the incredibly distant past, as though a god had taken records of all that was happening and engraved them for our present enlightenment. Time may have stood almost still, and only minor changes have taken place on the surface of the moon, since a period when there was no life on earth.

If this proves to be correct, it is so extraordinary a fact as to merit a moment of surrender to its poetry and mystery. Is this not moon-magic as weird as any assumed by primitive, ancient, or medieval man? We need not rely on risky mathematical extrapolations or on straining our vision of the distant limits of the universe, for nature has placed, conveniently close to our telescopes and nicely illuminated, a hieroglyphic script on an almost unchanging silver parchment of pumice or lava telling the story of the cosmic processes that formed the solar system. We have only to learn the moon-language and we may know what was happening before there was life and before our own terrestrial continents and seas came into existence. Will any human being ever again dare to think that cosmologists are abstract dreamers or that the secrets of the cosmic past are beyond the reach of the human mind?

The moon has always helped to shape the dreaming imagination of man, as the God or Goddess of the Night: SIN-NANNAR, SELENE, PASIPHAË, LUNA, DIANA. The moon in all its phases—as new moon, waxing crescent moon, full moon, and waning and dying moon—provided the earliest and the most powerful symbols of the rhythm of growth, and of the swelling and shrinking of man and woman in the course of procreation. The moon was water and all that liquid means to man. Listen to Plutarch:

> For other philosophers say "Whilst the light which flows from the Moon is of a moistening and prolific nature, and consequently very suitable to the generation of animals, and to the vegetation of plants—the Sun on the other hand, flaming out a more intense heat, scorches and dries up the young and tender plant, renders a great part of the earth uninhabitable and frequently gets the better of the moon itself."

Sin, the moon-god of Ur, linked the rhythms of plant life with the cycles of the heavens. The moon determined both the quality of time, the new moon being auspicious and the waning moon evil, and its quantitative measure, for "moon" and "month" stem from a common root. An Egyptian lunar god,

Thoth the "measurer," carried the scribe's pen and palette and was regarded as the inventor of the exact sciences. Primitive man threw his spear at the moon to hasten the future, and twentieth-century man pores over its surface to see into the past.

It is surprising, and yet appropriate, that the moon should now prove to be the most reliable record of the preorganic past, for the moon always provided the supreme symbol of paradox, contradiction, ambiguity, and complementarity, and today it achieves this once again. By waxing and waning it expressed growth and decay, and yet it is the only object of unchanging characteristic structure that has been seen by every member of the human species with the power of vision. The moon looks bright, and yet it reflects only 7 per cent of the light falling on it. It is seen by all, and yet only half seen. To the Germans, Slavs, and early Anglo-Saxons the moon is male, yet it is female in the classical myths, for the Latins and the English. And yet, in spite of all this two-sidedness, "moon-struck" may one day mean "struck by the unquestionable objectivity of history."

Galileo saw the moon through a telescope in 1609; Kepler wrote a dream fantasy on the moon in 1610; in 1946 radar signals were received by reflection from the moon 2½ seconds after emission; if a human being ever reaches the moon and sends signals back it will be—I suggest—at least some hundreds of years hence.

The moon is 240,000 miles away, has a diameter of 2,160 miles (about a quarter of the earth), a mass about $\frac{1}{80}$ of the earth, and rotates so as to conceal one side, though owing to the elliptical shape of her orbit we see altogether 59 per cent of the surface. And the moon's surface is unique: there are no definite shapes or forms in the known universe which are certainly as old as many of the markings on the moon. Everything else is guessing, speculation, abstraction, and calculation, beside the unchallengeably real and objective and highly significant forms on the face of the moon.

There are dark *maria,* or frozen seas of lava, lighter scattered mountains, and thousands of circular craters. These craters range in size from little crater pits one mile in diameter, a thousand feet deep, with a rim raised three hundred feet, to great crater rings twenty miles in diameter with depth and height in proportion, and finally to the immense mountain-walled enclosures, of which the largest is the Mare Imbria, seven hundred miles in diameter, with its surrounding ring of mountains broken by occasional outlets.

When the astronomer examines these through the most powerful telescopes he can distinguish objects which are only a few hundred feet apart. A crack, a ridge, or a rock a mere hundred yards across, formed perhaps a thousand million years ago, is still there for our inspection.

But there is another feature, perhaps even more intriguing than the circular craters: great systems of rays, or lines of lighter or darker color, extending usually in straight lines or great circles for distances of up to one thousand miles, each system radiating either from one major crater or from a point in a great sea. Many of these highly symmetrical radiating systems cut right through the mountains lying in their path, jump valleys, and tolerate no obstruction.

Once again we find ourselves making deductions from present forms to past history, and our scarcely conscious mental inferences run something like this:

1. The moon's surface must have been formed at some time or times in the past.

2. Single well-defined forms, such as craters and rays, must each have been formed at one definite time.

3. Very simple symmetrical systems, such as a system of straight rays radiating from one point, must have had simple causes, such as one explosion at that point.

4. The best defined forms are those which have been least disturbed and are still most nearly in their original state.

5. If two definite forms intersect, such as a ray and a mountain or a ray and a crater, the form which is perfect and unbroken must be the later of the two.

For example immense ray systems radiate from the crater Copernicus and from a point in the Mare Imbrium, and these rays override and are imposed on the mountain ranges around. But the Mare Imbrium is not itself scarred. The mountains in question must have been formed first, the explosions which created the rays came next, and the solid surface of the Mare Imbrium was formed after the explosion which created the rays diverging from a point in it. There are catches in such reasoning, but if used with care it enables not the quantitative chronology, but the temporal sequence of many of the formative processes to be reliably inferred.

Anyone who lets his imagination play on the surface of the moon must surely become moon-struck, impressed by the unexpectedness and objectivity of these markings, their startling symmetry, and unmistakably meaningful intersections. No metaphysician, philosopher, mathematician, or other speculative mind could dream up those markings with their positive message of a specific history. They prove the astonishing reality of the past, of definite events occurring in a particular sequence. The surprised delight one experiences on recognizing a fan of great valleys, for that is what the rays are, is profoundly scientific; the fact of one's surprise is the criterion of its objectivity; and the simplicity and symmetry are evidence of a single definite historical event. One did not expect those perfectly radiating valleys. They are quite unlike both the vague processes which certain pseudo-historians seek to impose on the human past and the neatly symbolic dreams of some persons undergoing psychoanalysis, which being expected and desired fulfill our wishes by appearing. The moon is today as fertile a field for science as it has been in the past for symbol and poetry.

There is as yet no acceptable theory of the origin of the solar system, and there may be none for some time. During the last thirty years some four different doctrines have held the field in turn, but no one has lasted long, and the latest is too new to be trusted. Yet the outlines of the story are slowly growing clear.

At one time there were no sun and planets, but only an unstable rotating disk or vortex of cosmic gas and dust and radiation with a major condensation at the center. Then some unknown factor—the close approach of another star, the development of turbulent eddies, or gravitational condensation around special points in the unstable disk—led to the formation of the embryonic planets. The moon and the earth were probably formed simultaneously, possibly by accretion and condensation of the cosmic dust.

When the major ray systems are examined under powerful telescopes they are seen to consist of scars, cracks, or valleys up to thirty miles wide. The conclusion is unavoidable that they were formed by immense explosions at the center which flung great blocks of solid material, up to several miles in diameter, flying in all directions at high velocities. Each ray system is the record of a single event which may have taken place between one and three thousand million years ago, but itself occupied only a few minutes. I confess that these rays and this extraordinary inference from them impress me more than even the perfection of Saturn's rings, for they reveal the queer versatility and definiteness of cosmic history. A process on the moon taking anything up to half an hour left its record unchanged through all the ages while life was emerging and evolving on the earth, until man arrived to examine it. There is nothing vague or universal in the strange definiteness of this story: at a certain place on the moon at a certain historical moment this particular thing happened and left its reliable record, and man evolves and interprets it, no doubt making mistakes at first but ultimately acquiring valid knowledge.

The reliability of the knowledge that will ultimately be reached is suggested by the fact that today the best specialists do not yet claim to be quite certain what created the explosions that formed the craters and rays. They may have been due either to internal chemical or volcanic processes or, more probably in the case of most of them, to the impact of immense meteors or planetesimals falling onto the moon's surface. At the present moment both theories have their supporters, and it may be that both causes were at work. But if the solar system was formed by progressive condensation of a cloud of cosmic dust then it seems inevitable that there must have been at some stage a period of cosmic storm or bombardment, when the embryonic planets were subjected to the impact of showers composed of myriads of dust particles with occasional larger objects and perhaps rare meteorites weighing up to one thousand tons, and still rarer occasions when a solid mass of ten-to-one-hundred-mile radius may have fallen onto the growing planet. In this strange manner the stage was prepared for life.

Some of the moon's larger craters were probably formed in this way, and the maria either by the heat generated under the impact of major meteorites or by molten material erupting from the interior. But in either case it is almost certain that the moon's surface reached a final equilibrium state earlier than did the surface of the earth. For on the moon there is no moisture or atmosphere and there has been therefore no erosion of the surface, apart from some wear and tear due to temperature changes, whereas on the earth the geological processes which formed the continents and seas and the erosion due to moisture have eliminated nearly all traces of the period of cosmic bombardment. While there is evidence of continuous change on the earth since its surface was formed, there is little sign of alterations on the surface of the moon since the great bombardment or the volcanic explosions which scarred its face. When we examine the moon we are probably studying not merely the birth of the solar system but almost the earliest days of the history of the universe of which we have any reliable direct or indirect record.

Beyond a certain point numbers evoke no sense of magnitude, but the following table shows that the moon is about five hundred thousand times older than the Great Pyramid, and perhaps five times older than the oldest fossil organism:

	Years ago *
	(approximately)
"The Origin of the Universe," or earliest calculable date	4,000,000,000
Formation of sun, earth, moon	3,000,000,000
Emergence of first cellular organisms, probably between	1,500,000,000
and	1,000,000,000
Oldest fossil organisms more than	500,000,000
First amphibia, trees	270,000,000
First birds	140,000,000
Monkeys, apes	40,000,000
Higher anthropoid ape	15,000,000
[A large gap occurs here in the dates already estimated]	
Primitive species of Homo uses stone chips but does not fabricate tools	700,000
Tool-making species of Homo, from	500,000
Complex stone-flaking techniques	250,000
Homo sapiens emerging, say from	400,000
to	100,000
Various types of Homo probably develop articulated speech, between	200,000
and	100,000
Earliest known agricultural civilizations	7,000
Great Pyramid built	4,600

The purpose of this chronological table is not to suggest accurate timing, but to evoke a sense of the historical sequence of these processes, and of the strangeness and grandeur of the story. Moreover the details are progressively falling into place.

* Most of the items refer not to definite events but to extended periods, and some of the dates may be in error by ±20 per cent. Part of this table is taken from Zeuner (see Follow Up).

Universal chronology is now becoming an exact science, whereas to Newton "chronology" merely meant the dating of the Ancient Kingdoms.

A mere hundred years ago Darwin could treat the problem of the origin of life as lying outside his field of interest. But today the time when the first living cells appeared is already rather closely bracketed: it must have been considerably later than three thousand million years ago, and much earlier than five hundred million years ago. Though the exact manner of the emergence of life is still uncertain, the frame has been set, the chronology is growing clear, and the precise chemical and thermal conditions on the earth at different periods are already being estimated.

No one who is familiar with the recent work on "dating the past" * can doubt the ability of science to penetrate the history of the universe and of the evolution of life on this planet. For a comprehensive sequence of dates extending from this century back to the formation of the oldest rocks on this planet can now be reliably determined to within an error of 10 or 20 per cent by comparing the results obtained by several independent methods whose ranges overlap.

The most important of these methods are: counting tree rings (back to about 1,000 B.C.), counting annual layers of sediment in clays and sands called "varves" (back to fifteen thousand years ago); astronomical and geological estimates for the Ice Age (reaching back about one million years); and radioactivity techniques (providing dates up to three thousand million years back).

These methods provide a chain of overlapping and mutually supporting estimates which justify the chronology given in the

* See book with this title referred to in Follow Up.

table. Three thousand million years ago there can have been no organisms and no primitive form of life in the vortex of atomic dust and radiation which was in process of forming the solar system. Five hundred million years back fairly complex organisms were already leaving fossil records.

If we exclude as empty or valueless such hypotheses as arbitrary divine creation, continuous spontaneous generation, and arrival of spores from cosmic distances, we are left with only one acceptable working assumption regarding the appearance of organisms on the earth: they must be the result of a continuous step-by-step development of increasingly complex systems of chemical processes, leading ultimately to the establishment of living cells capable of self-reproduction.

If this assumption proves correct, as I believe it must, and leads to a detailed theory of the origin of life on this planet, it does not necessarily follow either that a living organism can be synthesized by human skill in an hour or a year or that contemporary physical or chemical theories are adequate to provide a theory of biological organization.

But this assumption does imply that while time stood almost still on the neighboring moon natural processes on this planet molded out of inert materials the human person and his brain. Could versatility go further, or offer a greater challenge to the human intellect and imagination? There on the moon hardly any change at all, here on the earth the furthest exploitation of formative tendencies latent in natural law.

VI.

Invisible Structure

*A progressive evolution of micro structures
of increasing subtlety has taken place on the
earth*

IN THE vast tapestry of human experience there is nothing
so extraordinary, or so challenging to human understanding,
as the process by which in a partly haphazard world inorganic
materials exploited their own laws so that organisms came into
existence, and a cumulative process was started that led to the
appearance of man. This process is unique in at least two
respects: we do not know that it has happened anywhere other
than on this earth; and it made greater demands on natural
law than any other known process. To achieve this result
inorganic systems must, by a remarkable constellation of conditions, have somehow been enabled to realize potentialities
hitherto latent in nature. Life must have arisen by the exploitation of hidden aspects of laws already in operation.

In a *fully* haphazard world so subtle and sensitive a process
could never have come to fruition. The transition from inanimate systems to stable self-reproducing organisms was possible because, and only because, there were factors in natural

law which acted as nurse and guardian to the hypersensitive transitional and embryonic forms of life. These natural protectors, without which the universe could never have reached its flower in life and man, were the formative aspects of universal laws not yet discovered. Common sense and compelling logic lead to the conclusion that only if natural processes are inherently formative could life itself ever be formed. Without a bias toward form or pattern or regularity organisms could not have come into existence, or if by some weird chance they had come into existence they could not have survived.

To understand this better and to gain an inkling of the path from the inorganic toward the organic we have to make a radical adjustment in our ordinary point of view. An effort of intellectual imagination is necessary to visualize the micro world of atoms, molecules, genes, and enzymes.* For though these minute structures are from one point of view an expression of atomism, the laws they obey are pattern-formative rather than atomic. Thus by a peculiar irony when we go to the atoms to look for atomic laws, we find them often obeying pattern laws. It is down in the atomic world that we must take the first lessons in the laws of pattern and of life.

Human life enjoys its own natural scale of space and time. The hand can conveniently grasp objects of a certain size, and the eye discriminate to a certain degree of fineness and no further. So man began by making his own body the measure of everything. At that time he saw himself as the microcosm contrasted with the macrocosm of the earth and the stars.

The human body will always provide the natural measure

---

* Genes and enzymes are the characteristic structural units and catalysts in any organism, minute structures which guide and facilitate biochemical transformations.

for man as a person, and the rhythms of time which extend from the heartbeat to six score years and ten will always be of the most direct significance. But man is also an explorer who goes far in transcending his personal situation in order to open up new realms. And in some of these he has to accustom himself to strangely impersonal measures of space and time.

The vast distances and times of the cosmologist are in some degree objective and meaningful, though they may require revision and reinterpretation as knowledge advances. Yet they will tend to lack concrete significance for the ordinary person until they can be brought nearer home than the nebulae, which are far away not merely in distance but also in terms of ordinary experience. That is why the moon may prove a useful model for cosmic history: it brings past time near, on our own terrestrial scale of space.

We have now to consider another realm which man has discovered, perhaps easier to visualize than the cosmic, because in one sense it lies literally *within* our own experience. Not that the ordinary person in going about his daily life ever experiences this realm. But one only has to take a small part of any ordinary object and magnify it enough and one has entered this universe of fine structure that is normally invisible.

Suppose that you and I are together, and that you suddenly get smaller and smaller, like Alice in Wonderland, leaving human measures far behind until you reach what we may call the atomic scale or natural measure of fundamental structure. By then to your eyes my body will have became a macrocosm, a stupendous universe, highly subtle and complex, crowded with patterns and processes, small and large, and yet clearly all subject to some overriding ordering principle since the business of this macrocosm is carried on without confusion.

The human mind has long been haunted by a vision of worlds within worlds, of a universe hidden from the sight of man within the smallest grain. Early Indian and Greek thinkers suspected the existence of minute atoms, and the poetic imagination has seen infinity within a single point. Here is Pascal, writing in the middle of the seventeenth century:

> But to show him another prodigy equally astonishing, let him examine the most delicate things he knows. Let a mite be given him, with its minute body and parts incomparably more minute, limbs with their joints, veins in the limbs, blood in the veins, humours in the blood, drops in the humours, vapours in the drops. Dividing these last things again, let him exhaust his powers of conception, and let the last object at which he can arrive be now that of our discourse. Perhaps he will think that here is the smallest point in Nature. I will let him see therein a new abyss. I will paint for him not only the visible universe, but all that he can conceive of Nature's immensity in the womb of this abridged atom. Let him see therein an infinity of universes, each of which has its firmament, its planets, its earth, in the same proportions as in the visible world; in each earth animals, and in the last mites, in which he will find again all that the first had, finding still in these others the same thing without end and without cessation. Let him lose himself in wonders as amazing in their littleness as the others in their vastness. For who will not be astounded at the fact that our body, which a little ago was imperceptible in the universe, itself imperceptible in the bosom of the whole, is now a colossus, a world, or rather a whole, in respect of the nothingness which we cannot reach? *

* From *Pensées* of Pascal, translated by W. F. Trotter in Modern Library, Random House.

In one respect these speculations were right: there does truly exist a micro universe of structure. But in another respect they were wrong: the micro world does *not* mirror the macro. The training which the scientific mind has had in watching the heavens and studying the mechanics of ordinary bodies has made it difficult to recognize the very different laws of the atomic world.

Pascal saw organisms within organisms. The atomic physicists of 1910–20 saw a miniature solar system within the atom modeling the real one. The human imagination expects nature to repeat itself. But the facts are stranger and nature richer than our expectation. For new laws operate in the realm of micro structure, and there is every indication that they are very unlike any that have been discovered elsewhere.

Galileo, Kepler, and Newton mainly studied the interactions of two unchanging bodies, such as the sun and a planet or the earth and a stone. This was probably the best start for exact science, but the deep habits which they established have now to be overcome. For in the micro world no two entities ever dominate the situation, everywhere we have to deal with complex changing patterns composed of three, or thirty, or perhaps three thousand particles. The timeless cycles of planetary motion are quite unlike the pulsations of the complex organic units as they carry on their hereditary professions. Here science needs a more powerful language and calculus capable of describing the processes of complex patterns and structures.

It is a pity that during our spell of life on this planet we are not each granted one fantastic wish, such as the opportunity to enjoy another quality of experience for one brief hour. A man might choose to be a woman, or to be Akhenaton in his domestic circle, Socrates at the Banquet, Caesar first realizing his powers, Columbus reaching the American shore, Kepler discovering the planetary laws. . . . Or one might ask to experience space and time on another scale, in one hour to watch four thousand million years of cosmic history. But many exact scientists, the physicists, chemists, and biochemists who are victims of their own particular passion, would perhaps prefer to become a billion times smaller so that with their own eyes they could uncover all the secrets of ultimate structure at last become visible.

For ultimate structure must exist. If there is a grain of truth in atomism there must be last indivisible particles, final units which act as the markers of the root patterns of this universe. Such absolute assertions are treacherous, so let us add "within the physical world as we can know it with our present instruments and ideas." For the twentieth-century scientist there is a practical limit to fine structure; *we have already touched bottom.**** It is not our concern if someone in another century proves that it was a false bottom. Our task is to understand what is within reach, and we have struck a barrier that warns us that we shall be wasting our time if we try to get further— with our present methods.

Democritus knew that the doctrine of atomism would only work if there existed ultimate atoms of a definite size. And the twentieth-century physicist assumes that he has reached these ultimate units when he speaks of "fundamental particles." We have not yet attained full understanding of them, but in some

---

* This is a technical matter which cannot be explained here. But the evidence supporting this view lies in Heisenberg's indeterminacy principle, the fundamental character of such particles as electrons, protons, etc., and the universal role of such constants as the charge on the electron, Planck's constant, etc.

sense still to be clarified the neutrons, protons, and electrons of recent physics are the final units of this era of physics, the points or vertices of the basic patterns of nature. If only, like Alice, one could go down there within oneself and have a good look at a single cell inside the muscle or nerve of one's finger, or better still, to start at the bottom, examine the nuclei of the atoms, the shapes of the molecules, the working of the bio-catalysts, and all the still mysterious alchemy of life.

But we must be patient; the secrets are there awaiting discovery, they are of finite complexity, and many of them will be known during this century. Unless there is some cosmic disaster or the human race commits suicide the microcosm of structure invisible to the human eye is bound to yield up many of its secrets during the coming decades. For the optical micro-scope, X-rays, the electron microscope, and many other new techniques are already bringing much of this world of structure within the range of almost direct observation. During this century the grand inquiry into the nature of this universe has here reached a unique and peculiarly fertile phase: research is converging with unprecedented *élan* on the supreme problem of ultimate structure, and is confronted by one comprehensive and closely integrated task: the final unveiling of the structure of the nucleus, the atom, the molecule, and the operation of the biological units. For the next great wave of advances can scarcely avoid affecting all these realms together.

The problems here are of finite complexity. There are a finite number of chemically different atoms,* a limited num-ber of types of permanent particles (perhaps only one, the nucleons), and nuclear, atomic, and chemical processes can

---

* About ninety, though some nine hundred kinds of different atomic nuclei are known, of which six hundred have to be prepared artificially.

be broken up into combinations of a small number of elementary steps. There are, in other words, a finite number of different elementary unit processes, and everything that happens can be built up from these. Material atomism is facing a crisis, but an *atomism of process* is developing meantime. Everything happens by unit transformations and pulses. Once we understand how to describe these unit processes properly, the secrets both of the nucleus and of the biological units will not elude us much longer.

Moreover the realm of structure is spatially finite. There is only a limited range of spatial scale between the atomic nucleus and a small bacillus or a blood corpuscle. Expressed roughly and simplified for convenience the micro world is like this, in atomic units of length:

The Micro World

Size Ranges,
in *"Atomic Units"* *

[1/10,000	Atomic Nuclei]
1—10	Diameters of atoms and small molecules
10—100	Molecular structures determining specific biological properties
100—1,000	Biological units carrying these structures, *i.e.*, genes, enzymes, viruses, etc.
1,000—10,000	Diameter of small bacillus or small cell
10,000—100,000	Red blood corpuscle
1,000,000	Approximate limit of direct vision

The atomic nucleus is in brackets as there are reasons for thinking that in such very small regions "length" may lose its

* These units are taken as 5.10^{-9} cm. in this table.

ordinary meaning. If so, the nucleus may not belong in this table.

Notice that there is only a limited range of spatial scale between atoms and organisms. The diameter of a small living cell is only about a thousand times the size of a small molecule. Thus though there is room for an immense number of atoms in a single cell, there can only exist a restricted range of structural complexity. One may express this by saying that there are a relatively small number of successive domains of morphology, or structural arrangement, as one passes upward in scale: nuclei, atoms, small molecules, large molecules, micelles, cells, tissues, and finally multicellular organisms. The realm of invisible structure is not infinite, and the position of the structural scientists is like that of an army of explorers all converging on one point which can scarcely elude them.

Let us imagine that we have made ourselves so small that we can actually see this invisible structure. We should probably first notice some superb geometrical vistas, perfect perspectives in three dimensions made up of regular linear arrangements of identical units. These are like a three-dimensional orchard with rows of identical objects planted perfectly at regular intervals so that they make great linear arrays, in lines, and sheets, and solid square- or diamond-shaped patterns. Over here in this corner you can see a one-dimensional pattern, that is a linear molecule or a thread or fiber; and a two-dimensional grid, that is a little platelet of some laminar material; and the three-dimensional ones are crystals. These are all made up of regular *linear* arrangements of identical objects, each of the objects being either one atom or a group of atoms. The mathematician calls these arrangements "regular rectilinear point systems" or point lattices, and there is nothing in principle to prevent these lattices going on forever and filling up all space.

We would feel quite at home looking at these linear patterns, and for a good reason. Since the human species began to build permanent homes for itself and to mark out fields by using stretched cords it has been accustomed to linear repetitions of identical units. One brick is laid beside another to make a wall, and a length of cord is used to provide a repeating unit of distance in measuring the ground. The use of straight lines and of linear units is so deeply engrained in our way of thinking as almost to monopolize our attention.

However we might also see something else. Let me use the image of divinity once again. Imagine that you were the Creator setting out without any preconceived ideas to see what kind of a patterned universe you could make by arranging material points or particles in ordinary space. The rules of this creative play prescribe that every conceivable idea must be exploited, since it would be arbitrary to prevent any potential beauty of pattern from coming into existence. This implies that any principles of arrangement which you or I can imagine must have been conceived and used by the Creator.

Of course this universe may not have been formed according to such rules, but it is a fertile game since it encourages one to discover whether the Creator missed using any ideas one can think of. Sometimes that is hard to determine.

I will show you how the game works by giving an example that leaves the question open, and leads you, even without a degree in mathematics, right into the no man's land of science in front of the regular research teams. Remember we are down in the micro world gazing at all the patterns: points arranged in straight lines, in grids, in solid arrays, and . . . how else?

Playing the game well means setting the Creator a really

subtle problem, in retrospect. We must rebel against what has become commonplace and conjure up something new. We must not object to what seems very strange, that is just what we are looking for.

Why are all these well-known patterns *linear?* Surely because human dwelling houses are linear, and linear arrangements are therefore easiest for us to manipulate. But are all houses linear? The Eskimos build igloos without plane walls and roof, as one continuous hemispherical dome. This is possible because they do not use identical units or bricks but a continuous medium, snow, which can be molded to fill up any cracks. Have the Eskimos beaten the Creator? With the empty universe awaiting His space-filling touch did He not find some corner in which to employ *spherical* arrangements? Did the human preference for bricks, linear units, and Cartesian co-ordinates (length, breadth, and height) exist even before man, and somehow prevent the use of angles and spherical arrangements? The idea is absurd.

To the physicist structure usually means arrangements of points in space. Mathematicians accustomed to a linear world have assumed,* without realizing that it was an assumption, that only linear point arrangements interested the Creator. No wonder that some problems seem very difficult, such as nuclear structure where spherical arrangements may well play a role. It might be helpful for Departments of Nuclear Physics to take on an Eskimo mathematician,** accustomed to the task, so

* Until recently, see Follow Up.
** Or one might put this problem to a Russian mathematician: If Stalin had died as ruler of the whole earth leaving n sons, each of which distrusted every other equally so that they must all be kept as far apart as possible, how should his will have read? Where should the n sons build their n Kremlins? This is one example of a wide class of problems in the theory of spherical point arrangements.

bizarre to us but so everyday to him, of decorating the interior of an igloo with a number of identical sea shells arranged as regularly as possible. This might produce a renaissance of mathematics that can be visualized and of concrete physical theory: the study of Eskimo decoration or spherical point arrangements. You can try it for yourself: What is the "best" spherical arrangement of five (or seven, ten, eleven, etc.) points, how can they be distributed most uniformly on a sphere? And what does one mean when one says "most uniformly"?

Around 500 B.C. the Greeks knew the answer in the cases of four, six, eight, twelve, or twenty points: the vertices of the five regular solids, the tetrahedron, the octahedron, the cube, and two others. But until recently no one has investigated what you can do in other cases. I suspect that my Eskimo mathematician would say something like this:

"We only use four, six, eight, twelve, or twenty shells for igloo decoration—or rather half those numbers on a hemisphere —when a man wants to crawl into an igloo and die. These numbers are so perfect that we can't put an entrance tunnel anywhere, without disturbing the pattern. The volunteer for death uses one of these special numbers, decorates his shelter accordingly, and realizing its perfection is content to die.

"But for ordinary igloos we use almost any other number of shells, for then there is always an imperfection or gap somewhere in the pattern, and there we leave a channel for coming and going. This imperfection symbolizes the incompleteness of everything that is alive and can attract things to itself in seeking completion."

In this problem from a still uncreated branch of mathematics there is all the philosophy you can want. The Greeks studied perfection and exhausted the perfect and therefore essentially static spherical arrangements. We who have tried hard to be Greeks, but not being Greeks have failed miserably, have clung to ideals of static perfection, and are only now recognizing that the facts of process, of nuclear and atomic chemistry and of life, of necessity involve incomplete patterns trying to become complete. For the mere abstract idea of an incomplete pattern becomes alive of its own accord; it has a hole which must be filled and this sets going the whole machinery of process, the valencies of chemistry, and the attractions and repulsions of life.

This is one great secret of pattern philosophy: incomplete patterns possess their own inherent *élan*. The mathematical symbolism of patterns displays a tendency and movement of its own: toward completion. An Aristotelian formative tendency is a necessary consequence of incomplete patterns of Democritan atoms! Here is the marriage of atomism and holism: spherical structures seeking completion.

I suspect that if we could get our eyes in focus down in the micro universe, we would recognize within the linear arrays of the crystals some spherical patterns, integral arrangements not extending forever, but finite and arranged around one point. Moreover there are reasons for thinking that these may play some part in the nuclei and atoms, and that being smaller they were formed first, *before* the larger linear patterns.

Here we reach another idea of which there is little mention in the scientific literature until recently. If the universe began as a uniform atomic gas then a definite succession of steps must have been passed through between that initial state and the present condition of things. Along the main path of structural advance simpler structures must have *preceded* their more complex variants. In fact within the story of the evolution of the visible forms of organic species there must have taken place a more fundamental *progressive evolution of invisible structure*. This is one of the major new ideas resulting from the twentieth-century conception of structure. There is a necessary time sequence in the historical development of structure. Early in the history of the universe it is probable that no complex structures like molecules, crystals, or organic tissues were in existence. Structure is not a thing which comes into being arbitrarily, it must grow from simpler to more complex forms by a series of connected steps.

At the present time the physical and biological sciences are searching for a structural picture of all kinds of developmental processes, that is of the processes by which nuclei, atoms, molecules, crystals, and organic structures are formed from simpler units. This search for the inner mechanism of formative processes is leading toward a new comprehensive science of micro structure, whose main task will be the tracing of the story of the development of micro structure during the history of the universe and on this planet. From the cosmic dust to the pulsating texture of the human brain—that is a long journey however one measures it, but it has been accomplished and now we are using our brains to unravel it.

This grand theme of structural history is still in many respects a mystery. But it seems that it divides into a number of major steps which must have occurred in a definite succession:

1. It is useful to postulate a very early period in the history of the universe when there were present only *single particles* (say, nucleons), and no arrangements of them forming stable structures such as atoms, molecules, etc. This is a reasonable assumption because it is the simplest, and even if it proves to

be incorrect it meantime serves to start the description of the process. This may have been the state of affairs around four thousand million years ago.

But these single particles were unstable and tended to group themselves into more stable patterns, and so

2. Stable or relatively stable *chemical atoms* of various kinds came into existence in the gas cloud which was gradually forming the solar system. This may have been sometime between four and three thousand million years ago.

3. Some of these atomic patterns of particles were incomplete and therefore unstable, and combined to form relatively stable molecules, such as minerals, crystals, and salts in aqueous solution. This represents the Azoic or last stage of the preorganic phase of the earth's history, and it may have opened some three thousand million years back.

4. Then, possibly between two and one and a half thousand million years ago, there appeared very small regions of *complex regularly pulsating (? protein) structures* which were autocatalytic, that is tended to spread their structural pattern by a process of biosynthesis in the course of pulsating. This is a transitional stage of precellular life.

5. These regions of primitive organic pulsation were then stabilized within an enclosing membrane, and the dominant catalytic structures (genes) concentrated in an inner region (the nucleus), thus forming self-reproducing *living cells*. This stage may be put somewhat more than one thousand million years ago.

6. Multicellular organisms appeared, say eight hundred million years ago.

7. By a long process of evolution by natural selection the

highly subtle structure of the cerebral cortex of Homo sapiens developed, about three hundred thousand years ago.

The new historical science of micro morphology may be expected to clarify these steps and to determine their dates more reliably. But it will probably not admit any historical or theoretical boundary between the nonliving and the living. For the step-by-step development of increasing complexity of structure need imply no temporal or logical discontinuity. One unbroken continuity of structural process, or multiple branching threads of such continuity, must lead from some of the earliest chemical compounds at stage 3, through the first protoorganisms, the first cells, and the entire evolutionary process to each human being alive today. The individual atomic particles and chemical atoms may have passed to and fro without restraint, but there must have been *continuity of developing pattern* over more than a thousand million years. This is an astounding fact, which only a natural philosophy of form and pattern can recognize with its manifold consequences.

For if we accept this continuity of pattern in all earnestness, as a profound fact of natural history, it follows that we are the offspring of this continuity. This has unexpected consequences.

From 1600 until today exact science has rested on quantitative principles of a kind which possess no direct relevance to the more significant aspects of our experience as human persons. For example, the inverse square law of force and the mathematical theory of gravitation shed no light on the personal life, on human relations, or the general characteristics of organisms.

Yet if the continuity of pattern is valid there must exist

scientific principles of form and structure which have guided the entire process and must appear, though possibly disguised, in many realms, inorganic, organic, personal, and social. We can look forward to a unifying philosophy of form, displaying wherein we are one with all nature and wherein we are uniquely human.

This philosophy may not lie very far ahead, and its formulation may be eased by anticipation. For one can already recognize some rules which seem to be widely, though perhaps not universally, applicable. Since at this stage they can only be expressed vaguely, without specifying the exact conditions under which they are valid, they are certainly not yet scientific. But they may be on the way to become that. Here are some of these rules, supported by examples.

Stable structures are the end-states of processes and serve as records of them.

It is probably true that all structures were formed sometime, and examples of this principle are everywhere. The marks on the moon are records of very early events, and the lines on a human face tell of recurrent emotions.

Stable structures dominate the picture, since the unstable disappear.

Because of this human thought tends to overemphasize stability. There are more names for stable entities than for modes of transformation. The morale of an army, as of any human community, is sustained by that fact that grave casualties are quickly out of sight.

Incomplete structures are in some degree unstable, and tend either to complete themselves or to disintegrate.

Some chemical atoms are unstable; this may be because their internal structural pattern is incomplete. Other atoms are chemically active and tend to combine, possibly because their patterns are incomplete. Male and female are incomplete and seek to overcome this by coupling.

The presence of a particular structure facilitates the formation of similar, complementary, or identical structures, one model (or two complementary models) serving to produce countless progeny.

Organic catalysts, organisms, human beings, and ideas tend to reproduce themselves by facilitating the repetition of the process which formed them. Existing patterns tend to be extended.

Complex structures evolve from simpler ones.

In the evolution of invisible structure, in the evolution of organic species, and in many human activities there is a necessary succession from simpler to more complex forms. Development toward complexity occurs in a continuous sequence of relatively simple steps.

Complex structures display a tendency to interact selectively with closely similar structures.

Selective affinity and toxicity. Lock and key effects. High specificity in organic processes. Human allergy and selectiveness.

The properties of complex structures often depend on the character of the structural pattern rather than on the individuality of the units that make up the pattern.

Different physical systems are composed of the same fundamental particles, but arranged differently. The difference between one organism and another is due to a different molecular arrangement of similar atoms in the characteristic units (genes, etc.). The tradition of a community may persist even though no individual ever remains in it for long and every individual is different from every other.

These rules of structure should not be regarded as more than useful working assumptions which become reliable only when one can specify where they fail. But they provide glimpses of the vistas which will be opened by a science of form. They may develop into the axioms of a way of thinking which will reveal the parallels between all realms in which structure is at work.

It is for mathematical logic and exact science to clarify these rules and to establish an elegant philosophy of form easy upon the understanding.

VII.

What Is Life?

*"Life" may be regarded as the spreading of a
pattern as it pulsates*

WHAT is life? There are some who think it better that man
should never know the answer, lest he misuse the knowledge.
But human vitality expresses itself in the desire for understanding and will not be denied.

Others believe that man can never know the answer, because
they regard life as the expression of some nonrational power
which is beyond the comprehension of the intellect. Perhaps
they are misled by the magical aura of the word, "life."

And there are still others who suggest that the question is
meaningless, since there is no one condition or property that
can be described as "life." This may prove to be correct, but
here we are not concerned with final precision. Our aim is to
discover how the conception of form and structure can throw
light on living systems.

It is not too soon to attempt this. For the experimental
advance toward the problem of life is already in full swing.
The area of operations has been surveyed, the equipment
prepared, the converging routes have been reconnoitered, and

radical clarification should be achieved shortly, provided that the necessary *ideas* can be found. That is where disciplined speculation about form and structure comes in.

What is required is the development of ideas that can guide the crucial experiments and reveal the elegant order concealed in a myriad facts. Such ideas come best by relaxation and contemplation. If some of those who are now keenly pursuing their favorite techniques in logic, physics, biochemistry, anatomy, physiology, and psychology could be granted a year's leave of absence,* and allowed to dream and meditate, asking themselves only *what kind of idea is required to explain organic properties,* ten years' work might be done in one. For the greatest obstacle to advance is the fierce pressure of work along old ruts. Those scientists who feel that they have it in them might rebel, and cultivate patience, reverie, and the sense that it is important not merely to have a fixed purpose and to drive at it, but also to be an efficient instrument of the imaginative process. For the scientist who is impatient will be too tense to be able to surrender the misconceptions which persist in contemporary thought and prevent the emergence of a new clarity.

Many of those who consider that science is now close to fundamental discoveries regarding organisms none the less believe that the answer to the question "What is life?" will be so abstract, technical, and complicated as to have no significance for the average educated person. It may not be so. The specialist has always been the worst judge on such matters, for he is unconsciously prejudiced. The most fertile advances are simple. It is in the cards that the schoolboy and girl of the last

* Two recent reports in the United States have stressed the danger of overemphasis on group research, there being no substitute for the single human mind following independent lines of scholarly research.

decades of this century will understand how certain special arrangements of chemical constituents possess the properties of organisms, in accordance with fundamental principles that have not yet been discovered.

Expressed in scientific terms our question means: "What are the characteristics of living systems and what relation do they bear to those of inanimate systems, and of any transitional forms?" And this is almost the same as asking: What are the basic laws of physics, of biology, and of any no man's land between, and how are they related?

If I am not mistaken there will be chairs in all progressive universities expounding the answers to these questions before the century is out. For the benefit of the faculty boards who will have to make the appointments I suggest they should be called chairs in *The Theory of Structural Transformations,* for properly understood nearly all the processes of physics and biology and of the no man's land are structural transformations.

Do not imagine that this will be an abstract science without very concrete applications. I can imagine in fifty years' time the manipulation of big organic molecules and the step-by-step building of protein in a living condition. This might be done by using as micro manipulators not a pair of forceps but a system of focused electric and magnetic fields undergoing cyclic changes, which would guide the atomic groups together to form the pulsating molecular fabric which is what we call "protein in the functional state." That would be halfway to making a living cell.

But this is the 1950's. Have we got beyond the stage when the idea of a fusion of physics and biology in a comprehensive structural science still provokes the old nineteenth-century misunderstandings and protests: "You don't propose to reduce life to mere mechanism, do you? That would make man a machine!"

The idea of the objective study of structural relationships in all realms is several decades old, but it may be that these old prejudices are still about. There is nothing "mere" and little that is "mechanical" about the emerging structural science. "Mechanics" should mean the classical mechanics of machines, clockworks, elastic jellies, and chaotic swarms of hard spherical corpuscles, and ideas of that kind lost all claim to be generally valid around the opening of the century. Two demolition gangs under Max Planck and Albert Einstein broke up the old mechanical foundations,* and they did it so effectively that any future rebuilding will have to be done on a much deeper basis. The new structural science will have to build on ideas that bear little relation either to machines made of solid parts or to chaotic swarms of particles. There is no fear of any valid science in the future "reducing man to a machine." Man is man, and a machine is a machine, and the task of science is to discover their differences as well as their similarities. In any case it is now clear that "mechanical" ideas in the strict traditional sense only apply to highly restricted parts of the whole realm of natural phenomena. We may be on the eve of a structural explanation of life, but it will not be "mechanical" or "mechanistic" in the original senses of these terms.

The science of structural transformations and changing forms will be as rich and varied as are the facts themselves. And the "facts" cover all that the scientists and their techniques of observation are capable of recognizing at the present

* Using as explosives two natural constants, h, Planck's constant, and c, the velocity of light, respectively.

time, purified of misconceptions. The aim is to discover the relatedness of everything, and this aim is colored by only one prejudice: the conviction that beneath apparent complexity a simple order awaits discovery or can be invented.

Fifty years ago a biologist might have said that biological systems are distinguished by the fact that certain organic properties appear to be imposed on the chemical constituents. For example, he might have emphasized growth, differentiation, reproduction, functional cycles, irritability, metabolism, and self-regulation so that the organism maintains and reproduces itself. A contemporary biologist might rearrange and rename these properties, and he would probably add specificity (highly selective properties), self-reproducing units, unit steps of chemical transformation each involving small energy, and modifications providing the basis of learning and memory.

Around 1920–30 under the stimulus of Mendelism and of the developing science of biochemistry biologists began to realize that many of the properties of growing and adult organisms are partly due to the presence, both in the original fertilized egg and in every cell that arises from it, of minute chemical units probably of protein,* possessing a highly specific structure. These units are the "molecules" of biological structure, the smallest parts which still retain the specific biological properties.

This was an important step in the development of structural thought generally: gross properties, observable to the naked eye, were ascribed to the presence of tiny structures of molecular scale. It was realized that the characteristic properties of a given family, species, or sex of organism—such as its external form, internal anatomy, biochemistry, and physiology

* Or nucleic acid.

—in so far as these differed from those of another family, species, or sex, are partly traceable to the characteristic physical structure of ultimate units of nucleic acid or protein, such as the hereditary genes in the chromosomes of the cell nucleus, the enzymes in the cell cytoplasm, and any new types of gene or enzyme formed by mutation or otherwise.

These ultimate biological units of specific structure can be regarded as the organic bricks from which the organism is built, or the organic prototypes on which it is modeled, but both of these images are too passive. For the genes and enzymes are active factors * which, by their own reproduction in part or whole, do much to control all the developmental and functional processes of the organism. These units are not static models, molds, or templates, they are pulsating printing presses which operate by partial or complete self-reproduction of their own patterns. They are in fact the ultimate "proto-organisms-within-the-cell," and they fill the cell with their progeny until it is forced to divide. But these units are themselves without means of nourishment and defense, like viruses they are normally only stable and can only do their work within a living cell.

Physical atomism enjoyed its greatest success from 1880 to 1930. This new biological atomism developed somewhat later, from 1900 onward. But while a crisis of physical atomism began around 1932, it is probable that biological atomism will flourish for long. For the biological units are not independent static units, but pulsating structures kept going by a changing environment. The physical atoms are in difficulty today be-

* They probably pulsate when doing their job, like every organ in an organism, and biosynthesis may occur in pulses or waves. Recent research has emphasized the importance of structural pulsations in living systems of all sizes.

cause they claimed too much individuality and their patterns and interactions were unduly neglected, whereas the biological units are conceived as part of a wider organic environment and subject to its influences.

These biological units possess a property which is so un-mechanical as to show that we have entered a new realm. In suitable environments they are capable of continuing identical self-multiplication, the autocatalytic property of bringing about the extension or reproduction of a part or the whole of their own structural pattern. Thus reproduction is a characteristic not merely of organisms as a whole, but also of these minute proto-organisms. Moreover nearly all growth or increase of organic tissue is due to the self-duplication of these units, which impress their pattern on the digested foodstuffs and so assimilate the new material to the particular organism and cell. Each organism, and indeed each cell, synthesizes its own specific types of protein, and it does so through this catalytic and pattern-reproducing property of its characteristic structural units.

Further, the high stability of the properties of each species through thousands or millions of years or generations must be traced to these protein units, which either step-by-step or in a single process reproduce themselves *identically,* at least under normal conditions, if we neglect the relatively rare miscarriages, mutations, and any other possible changes.

Wherever cell division, growth, or repair is taking place in an organism these units are at work reproducing themselves. Their reproductive libido is inexhaustible, so long as the cell around them is alive. This unit reproductive activity is from the structural point of view the supreme process of life, and its recognition is one of the greatest biological discoveries of our time, or will be when we understand it better.

For though it can scarcely be doubted that something of this kind is happening, the details of the process are still a mystery. There is no accepted biological or biochemical explanation of this self-duplicating process, nor do we even know the exact structure of any particular type of protein unit, still less how a hereditary equipment of such structural units leads to the development of an adult plant or animal of a particular anatomical structure and external form. Genetics and embryology still lack a structural foundation.

I have used the phrase "specific structure." What does it mean? It does not directly concern *organic* species, but implies a *structural* species, a particular type of atomic arrangement, presumably rather complex like an intricate kind of chemical molecule. Many of the processes of organisms display a high degree of *specificity,* by which is implied that interactions either within the organism or between it and its environment, are highly selective, and occur differently or not at all if one of the interacting patterns is altered, even very little.

This is obvious in many realms. A drug may affect only one tissue; an established condition of immunity may be effective only in respect of one particular form of a disease, allergic forms of specific sensitivity are frequent, we are most of us powerfully attracted by a thing which is just right and cannot tolerate something else only slightly different. The examples are endless. All life displays highly selective affinity and often equally specific rebellion.

Inevitably one finds oneself thinking that all such processes must depend on a fine spatial correspondence, a close geometrical correlation, a structural similarity, like a lock and key but subtler and more dynamic, perhaps like a new kind of lock and key which both change their shape as the key is turned, or like an orchestra of instruments tuned to play in unison.

Most of the organic processes which display this kind of selective tuning do so because of the presence of these units of protein of highly complex individual structure. All natural proteins are built up of varied combinations of some twenty different types of chemical building unit (amino acid) and the variety of the different types of protein must be largely due to the myriad possible combinations of these twenty units. To each specific type of protein may correspond a definite sequence of these building units along a chain. Thus there is no difficulty in accounting for the almost innumerable organic species and their different types of protein and nucleo-protein. This specificity goes so far that if a graft has to be made to replace skin burned on a human, the skin must either be taken from another part of his own body or possibly from that of an identical twin, for the organism will regard any other human or nonhuman skin as in some degree alien and unacceptable, and the skin will wither after a week or a month.

Considerable advances have been made recently toward identifying the precise structure of various kinds of protein. The basic structural feature is a thread or chain of atomic groupings, and this chain is sometimes coiled into a three-dimensional spiral or helix, like a wire wound on a cylinder or a spiral staircase. Moreover these helical chains may be twisted round one another to form a compound helix, just as a rope may be made by twisting round one another a number of twisted cords. And in many organic processes, such as the action of muscles, these helical chains fold and unfold, coil up and uncoil, pulsating like a spiral concertina. It has been known for nearly two hundred years that the helix plays an extensive role in organic nature, for example in plants, and now it has been rediscovered in the fundamental structure of certain proteins.

From recent research in biochemistry and on the architecture of the living cell a tentative picture is emerging of the nature of biological organization. In a crystal the ordering of the system lies in the regular repetition of identical unchanging components, but this cannot be the clue to the arrangement of living material which always contains many contrasted components undergoing continual transformation. Biological organization is heterogeneous and "meta-stable" or liable to transformation.

It appears that in living material all the contrasted constituents tend to take up positions so that they fit together as nearly as possible, contiguous surfaces being covered by nearly complementary structures, and molecular units joining up with their neighbors, as far as possible. All the separate parts tend to fit themselves together as well as they can so that they form either regular arrangements or graded patterns. Identical constituents may join up to form either semirigid structures, like plant cell walls, or labile frameworks of protein chains capable of conveying influences from one point to another, like nerve cell walls. All the functioning regions of cells and of organisms seem to possess a labile but perpetually self-restoring skeleton structure of protein. And this sensitive self-repairing molecular fabric seems to hold the secret of biological organization, for it determines what happens where and when, and in what directions.

But the touch of life is still missing from this picture. It is not enough to assume that everything tries to fit together and so to reach the most stable positions. This would provide a static model, or a model seeking stability, whereas what we need is a *working* model, one that continues working.

The answer is that living structures are never static, they are continually undergoing cycles of change in which they expand and contract rhythmically. Our model must *pulsate!* The molecular fabric of the organism is perpetually undergoing rhythmic changes, cycles of expansion and contraction.

This provides a hint regarding the nature of biological organization, or the way in which the organic parts are ordered to make a living whole. The parts must be so fitted together that they can undergo pulsations simultaneously, or rather that the entire system can undergo a single co-ordinated pulsation. While a crystal is a static array of identical parts, an organism is a graded system of varied parts, fitting together well but not too well, so that they can pulsate together as a single system.

However this model of a complex system expanding and contracting rhythmically is still not complete. For the processes of living systems always display two aspects: a periodic or cyclic aspect which restores the normal state and leaves no net change, and a progressive or one-way aspect which results in the cumulative extension or multiplication of the organic patterns. In future when we speak of a "pulsation" in an organism, we shall mean a process which contains both these components, the cyclic and the one-way, the rhythmic expansion and contraction and some cumulative result. An organic pulsation is then a cycle of changes producing a cumulative result.

The heart in pulsating returns to its original state, or very nearly, but it maintains the circulation of the blood. The lungs undergo cycles, and in doing so maintain a perpetual intake of oxygen and discharge of carbon dioxide. The muscles contract and relax, but also do work. There is a systole and diastole, an up-phase and a down-phase, in the operation of every organ and structure, yet these cycles take place in a wider system where they produce a cumulative result. Fully developed organs change very little as they work, but their activity produces a one-way flow of energy and a displacement or transformation of material. Life is a stabilized mode of pulsation producing a cumulative result which is usually the extension of the characteristic organic pattern.

It is easy to picture how a structural pulsation may take place. Each organic structure has a most stable form, A, but can be deformed with absorption of energy and production of electric polarization to a less stable form, B. This upward recharging phase, $A \Longrightarrow B$, can only occur when the necessary energy is supplied, but the downward relaxation or depolarization phase, $B \Longrightarrow A$, can occur spontaneously and do work. Thus the cycle, $A \Longrightarrow B \Longrightarrow A$, leaves the functional structure unchanged, but there has been a flow of energy or material and some work has been done. As the cycles continue, the local structure or organ remains unchanged, yet a progressive and cumulative change is brought about in the organism as a whole, a change which normally tends to maintain or extend the organic pattern.

The suggestion is that the clue to biological organization lies in the fitting together of a number of different structures so that they can all be deformed and pulsate together, either simultaneously in unison or progressively in serial succession one after another along some functional channel. If this fitting together were too good, the system would jam and could not continue pulsating. Thus biological organization is from one point of view necessarily imperfect, incomplete, and unstable, though in the healthy organism it is always tending to improve and to restore itself after any disturbance.

It is for biological theory and experiment to show in detail how a given cycle of structural deformation and relaxation produces a particular one-way transformation. But the two aspects are always both present together and no organic phenomena can be interpreted in terms either of cycles alone or of one-way transformation alone. The physiology of functional cycles and the developmental theory (embryology, growth, learning, etc.) of the one-way transformations are complementary and mutually dependent. Here are some examples:

Cycle of a developed structure or organ	Cumulative one-way change
1. Pulsation of gene, synthetic enzyme, or synthetic cell.	Continual synthesis of new specific material, or spread of the structural pattern.
2. Pulsation (polarization and depolarization) of parts of a nerve cell wall.	One-way propagation of states of polarization, or spread of a pattern of polarization.
3. Pulsation of a muscle fiber.	One-way transfer of energy or displacement of material.
4. Pulsation of a neural ganglion (e.g., cerebral cortex) during cyclic changes of polarization.	Spread of existing patterns of polarization, *and formation of new patterns*.

Notice the inference for ourselves: a life which is composed of biologically balanced pulsations must comprise both functional cycles and progressive change, both periodicity and growth. The basic cycles of nutrition, waking, and sleeping, and work and play must be maintained but they should be complemented by an adequate degree of one-way process: of growth, reproduction, learning, and constructive or creative activities. With every breath, in and out, we grow older, but this can be complemented by a small residue of cumulative achievement.

If we examine the cumulative changes in the right-hand column which represent the results of the functional cycles, we notice that the result of most of them is to extend or spread the existing structural pattern. Thus we reach the idea that organic activity consists in the spreading of a structural pattern as it pulsates. At one end the organic structural units, the genes and enzymes, multiply their pattern as they pulsate; at the other the pulsating processes of thought lead to the spread of the patterns of thought.

We can understand this better if we contrast it with other summary descriptions of life:

"Life" is a name for all cyclic processes involving protein. This neglects the progressive aspect of all living processes, and also the fact that the cycles have the form of an integrated pulsation of a complex system.

"Life" consists of self-perpetuating patterns of chemical transformations. This also fails to reveal one essential feature, the complementary cyclic and cumulative components. Moreover life is not merely self-perpetuating, it is essentially self-extending or multiplicative.

"Life" is the production of order out of disorder under statistical laws. If "the production of order" is interpreted to

mean "the extension of specific patterns" this description is acceptable, but the emphasis on statistical laws is misleading, since pattern laws contain elements of form which must be assumed prior to the application of statistics.*

To these we now add another description: *"Life" is the spreading of a pattern as it pulsates.*

Perhaps this should be regarded as a program for research rather than a conclusion. For there is still much work to be done before the pulsations of organic structures are fully understood. Indeed a general theory of structure and form may be necessary, possibly using new principles and mathematical methods.

This idea of structural pulsations marks the convergence of two traditions: *atomism,* providing the precise structural skeleton of the new theory; and *holism,* contributing the *élan* of the system as a unit, the fitting together and the tendency toward the extension of form.

The conception of a generative principle or formative energy underlying all forms of life is very old and has recurred frequently through the centuries, clothing itself in the ideas of each period. We might go back to Aristotle, to Paracelsus, or to many others for examples. But here is Sir Thomas Browne, the English physician and author, writing in 1646: On

---

* "Statistical laws" should mean laws which determine only the probability and not the certainty of any particular event. But as pointed out before there are no *purely statistical* laws, for some aspects of events are certain, and not merely probable. For example, the three-dimensionality of ordinary space, the irreversibility of actual processes, and the presence of certain kinds of symmetry and asymmetry in crystals and organisms, are universal, and not merely probable properties. Indeed any statistical formulation presupposes the universal applicability of the particular quantities which it employs. "Statistics" can never be fundamental, for it is a method which can only be applied after some process of selection has been carried out.

crystals: ". . . which regular figuration hath made some opinion, it hath not its determination from circumscription, or as conforming to contingencies, but rather from a seminal root and formative principle of its own." On organisms: "The plastick or formative faculty, from matter apparently homogeneous and of a similar substance exciteth bones, membranes, veynes and arteries."

Today we can trace that seminal root or formative principle present both in crystals and in organisms to the structure of micro units which in extending their own minute patterns determine the macro pattern of the whole.

And this ancient analogy can be traced further, in a manner which no Pythagorean or Platonist could ever have conceived. For no one whose imagination was haunted by the idea of perfection could ever have reached the notion that *crystals and organisms grow mainly because they are imperfect.*

It is now established that the absolutely regular faces of a perfect crystal do not provide an adequate opportunity for rapid growth; it is only the presence of imperfections or dislocations which stimulates normal crystal growth. This is a nice example of the doctrine that perfection is static and that all process is initiated by discrepancies or differences.

Moreover the same is probably true of organisms. If all the constituent parts of an organism fitted either perfectly, or too nearly perfectly, or if the organism fitted its environment too closely, there would be no cause, or opportunity, or possibility of pulsation or transformation. It is only the imperfection of the fit, the difference between organism and environment, coupled with the perpetual tendency to improve the fit, that allows the working parts to work and makes them continue to work. Life is a movement toward a condition of perfect fit that can never be realized, while there is still life.

In this chapter I have offered as a provisional description of "life" the spreading of a specific structural pattern as it pulsates. But in the next chapter we shall see that this description, while it may be adequate to cover those organic phenomena in which an established species maintains itself, is too narrow to cover those processes in which *new* forms come into existence.

VIII.

What Is Man?

Only a science of formative processes can pro-
vide an adequate picture of man

WHAT is man? This may seem to be a somewhat morbid question. When vitality flows directly into action, there is no occasion to gaze into the distorting mirror of introspection. Yet the price of being human is the loss of thoughtless spontaneity of action, and in trying to understand existence men have always formed some image of themselves.

There does not appear to be any book which tells the story of man's changing views of himself. It is a remarkable process, this effort made by a part of nature to understand itself, and it raises peculiar questions that have not yet been clarified. Is the view which a thinker forms of man an expression partly of his own condition, and does it also influence that condition? And does it express what he is, or what he is not and would like to be? Are there features in every human person which he himself can never recognize, and is there perhaps in humanity some factor which can never be known to man through direct self-examination? Possibly in order to learn some last truths of human nature we shall have to learn to communicate with

121

another species. If we could read the mind of a cat, we might be surprised.

But as it is we have to help ourselves. We belong to a species every member of which shares this strangest of experiences: we are born unaware and only gradually become aware of the world and of ourselves. Moreover we do not arrive, nor did the species arrive, properly equipped with the kind of under- standing that would be natural to a species with our poten- tialities; we have had to collect such understanding as we could as we went along.

In a sense this is hardly fair, but fairness is irrelevant. It is a challenge. What is this fate of being human? What is man?

It is a pity that we cannot summon a Council of the Wise to pool views and see where we have got to in this adventure of self-discovery. We might hire a hall of two hundred seats, and lay down the name cards: Akhenaton, Zoroaster, twenty of the Greeks, Buddha, Jesus, Mahomet, Aquinas, Galileo, Newton, Freud. . . . How much did the wisest know? What does it all add up to? Is our understanding, such as it is, to be expressed in words, or perhaps in art?

But we cannot summon the dead, except through their records in script and art. And I cannot take you through the picture galleries or the buildings of the world and silently point to their answers to the question, What is Man? It might be better to do so, than to use words.* Intuitive awareness, ex- pressed in nonverbal form, comprises a greater range of ex- perience than the verbal and algebraic symbols of language and mathematics can yet convey. We feel and perceive much that we cannot say in words.

---

* Goethe wrote: "I should like to give up entirely the habit of speak- ing. There is something about it that is useless, idle, foppish. . . . I should like to speak like Nature, altogether in drawings."

But if we accept the limitations of linguistic expression, what ideas has man so far had about himself?

Here are some of the things he has thought and said. That man is:

The creation of a Divinity, serving Him. This may be interpreted literally or symbolically. We are certainly all part of something more extensive than our conscious selves.

A reincarnated spirit in course of development. This may extend the continuity of the individual mind too far.

An eternal mind imprisoned in a corruptible body. This implies a basic dualism which stifles deeper insight.

An instrument of a historical-social process. This neglects the role of the person in the social process.

A chemical system undergoing perpetual transformation. This is misleading for the known laws of chemistry cannot yet account for organic and human characteristics, and to achieve this they may require revision. In other words if "chemical" means "as known today" the description is wrong.

A conscious, autonomous, rational, and purposeful person. This covers only one aspect of man, the four epithets being only partially applicable. The unconscious source of all processes of growth, imagination, and creation is excluded.

A member of an organic species distinguished by a symbolizing faculty. Many animal species share this faculty.

A social organism marked by an unresolvable conflict between libido and reason. True only of certain individuals in certain cultures.

A primate with erect posture and the faculty of speech. True, but inadequate. We have to discover what precisely distinguishes human speech from the communication systems of animals.

If I had to act as secretary of that Council of the Wise, to take notes of their deliberations, and try to find the common truth underlying the differences of emphasis and contrasts of cultural background, I know what my report would be, assuming that Socrates had not achieved fresh insight in the intervening two thousand years, or Freud become a Platonist, or Nietzsche a Thomist. Anyhow I have no doubt regarding the language in which the report would have to be written. It would be that of a naturalistic or scientific mind trying to be receptive to all suggestions. For it is the task of a comprehensive science to understand everything that man has thought and the reason for his thinking it, and to discover the deeper truth compatible with all the contrasted glimpses of it. Only the objective methods of science can avoid the prejudices of introspection and of particular social situations.

The idea of a single report does not imply that it would seek to smooth out all differences and pay lip service to every mistake that the race has made in its stumbling search. The advance of thought involves hard renunciations, men fall by the wayside, ideas are eliminated, and ancient beauties destroyed by more powerful visions. The secretary would be of no use if he were not repudiated by the Council, and the rejection of his report would probably be the only point on which they would all agree. Unless of course they had each in the meantime moved toward a deeper understanding, as we now see it.

Is it foolish to attempt to summarize the wisdom of the ages and the knowledge of our own day? The power of words is great when they express a need. And the words which I shall employ express something for which I have needed to find words: the fusion of a conviction of human unity and of the unity of man and nature with an intense awareness of human differences, and of a recognition of boundless creative vitality with the knowledge of ceaseless frustration.

No single phrase can match everyone's experience. But if the Council of the Wise were here I would submit a summary report on man's understanding of himself and invite their criticism:

Man is the expression of universal, organic, social and personal formative tendencies in a world of accidents.

Man is the personal manifestation, the individual instrument, of forces much greater than his own conscious mind. Physical conditions, organic vitality, social history, and individual heredity and experience combine to make him a unique expression of universal processes which are essentially formative. But they operate in a universe where accidents continually distort them. I leave this seventeen-word description of man to the judgment of the future with confidence that the idea will be understood.

A great hierarchy of formative processes culminates in the creative powers of his unconscious mind, his imagination, and all his constructive activities.

Nevertheless this universal and organic tendency toward order and harmony operates in a world where chance circumstance and accident leave nothing alone. If value is placed in harmony rather than in the tendency toward it, and beauty is seen in the form rather than in the forming power, then tragic and bitter frustration is unavoidable.

Moreover implicit in this description of man is the sense, desperately needed today, that all human values arise from, and are expressions of, organic vitality operating in man. *Everything* that man is and experiences has to be understood as the expression of organic processes, life impulses, vitality in human forms.

Man is capable of awareness, reason, purpose, and self-control. But he possesses these attributes only in relation to particular aspects of special situations. They are subordinate to, and only operate properly as expressions of underlying vital impulses.

Man becomes aware of sharp divisions: mind/body, reason/ impulse, and inanimate/animate. But these categories are valid only in restricted contexts. One formative tendency and process underlies them all.

Man displays many apparently separate faculties and tendencies. But these are the differentiated expressions of one basic tendency and they remain potentially subject to its organizing control.

Perhaps the Council would not be unanimous in repudiating this summary. The ancients are too far away to share our experience, nevertheless Montaigne, Bruno, Spinoza, Shaftesbury, and Goethe—to name a few—and perhaps Albert Schweitzer, too, might not reject these suggestions. These men sought a unity between organic nature and human values, and in some degree identified human consciousness with organic vitality.

It seems that this is the indispensable step if present lesions are to be healed.

There are two sides to this restoration of unity between nature and human values. On the one hand the biological picture of man must be extended so as to include all human capacities and experiences. On the other hand the individual has to achieve a deeper awareness of himself so that all his values and aspirations are felt to be an integral expression of the organic vitality which he experiences in himself. In our time values which are not so understood are empty and hypocritical. All human values express man's aspiration to harmony and this is the flower of the organic tendency toward order. Tear the flower from the roots and the integrity has gone.

How much better if this could be left unsaid, for to express it inadequately is to wound the truth. But I have learned that some contemporaries do not regard their own values and aspirations as expressions of an underlying vitality, but as either separate from or in conflict with organic impulses. This is surely the expression of a morbid condition. In twentieth-century man any conscious factor which is not experienced as an expression of broader natural and organic principles must necessarily be distorted and in some degree lack integrity or sincerity. If the sense of the unity of the fully conscious and the less unconscious elements in man is lacking it means that somewhere, at some time, what was originally conscious has been inhibited and that some distortion has resulted.

Yet if the whole of human personality is viewed as the expression of organic processes an important question arises. Does this imply that all human faculties and modes of activity are to be accounted for as consequences of an evolutionary process which selected from random variations only those characteristics *which had actually proved their value in promoting human survival?* * In fact, are all the characteristics of Homo sapiens, as displayed, say, in this century the direct or indirect result of their proven reproductive value?

This is certainly not the case. As a species Homo sapiens is

---

* Or, more explicitly, in promoting the *reproduction* of the successive generations.

too young for the adaptive value of his distinguishing characteristics to be regarded as established. He may have survived no longer than other types of Homo which have disappeared, and he may still follow them. The entire history of man may be the expression of a set of unfavorable mutations, which on the short run appeared favorable as they enabled man to achieve mastery over all other species, but have still to demonstrate their lethal character.

On orthodox evolutionary theory that possibility must be regarded as still open. For Homo sapiens is manifestly not yet an adapted species with a stable pattern of life and an appropriate niche in the organic realm. Homo sapiens displays a destructive itch unique in the organic world, which may either be inherent in his heredity, or, as I consider more probable, be the expression of the fact that he is still struggling to establish the mode of life appropriate both to his hereditary potentialities and the environment as molded by himself. It may be a very long time indeed before this issue can be regarded as settled. In the meantime Homo sapiens is still biologically *sub judice*. His biological destiny is still open. Man is certainly unique, he possesses more capacity not only for good and evil, but also for maintaining or destroying himself, than any other species. But the balance is not yet drawn.

The trouble about man is that he is too vital, and his potentialities are too great, for him to find it easy to make good. He suffers from surplus capacities beyond the needs of biological survival and beyond anything that could be ascribed to natural selection of favorable characteristics. For this excess is potentially dangerous and continually threatens—and has so far upset—every temporarily achieved and quasi-stable mode of civilized life. Homo sapiens appears to be cursed and blessed

—for it is both—with a restlessness springing from still un-realized potentialities, far in excess of the degree of vitality that would be biologically appropriate or adaptively most advan-tageous.

A parallel from the subhuman world throws light on this situation. Recent zoological research has shown that many animal species display a surplus vitality or tension, an excess of nervous excitation and an overmotivation, relative to the requirements of normal instinctive behavior and of species survival. This surplus results in an overproduction of internal processes and of movements, with the consequence that activity patterns which were previously adaptive may become displaced and acquire a new structure and significance. The organic vitality which had previously expressed itself in stabilized in-stinctive patterns of activity, when it is in excess and unable to find its traditional expression, may shape novel forms of activity which cannot be the expression of proved adaptive value, for they may never have existed previously.

Tinbergen * gives an instructive example of these "displace-ment activities," which arise when an excess of nervous tension cannot find adequate release in a standard pattern of instinc-tive behavior either because of a conflict between two instincts or because the environment does not provide adequate op-portunities. In such a situation the pattern of activity belonging to one instinct may be employed out of context, so that it acquires a new significance, often becoming an expressive ritual revealing the internal condition of the organism.

Thus the male stickleback, when in an excited condi-tion with the attacking-drive and the escaping-drive nicely balanced, will adopt a threat posture (vertical with head

* See Follow Up.

down) which is actually a standard sand-digging movement displaced from its original use and employed as a means of releasing the tension arising from frustration of one or the balanced conflict of two instinctive drives. This new displacement activity is ritualized, and becomes a symbol of threat, though it is only a slight modification of a pattern which was originally part of the sand-digging and home-making pattern of activity.

The new threat ritual cannot be interpreted as the consequence of natural selection acting on random mutations, for it is the expression of an excess of tension shaping new patterns of activity during the life of the individual. There is no gene or set of genes determining this particular pattern of behavior, which is evoked only by a particular social situation. Such new modes of behavior are possible because the animal possesses a brain which has the capacity of shaping new patterns of behavior, and they cannot be regarded as necessarily of adaptive value. The brain of any adapted animal species has survived the process of evolutionary selection because it can and has ensured the survival of the species, but it also does other things in addition to that: it shapes new forms of behavior which may or may not prove advantageous, and it does this *before* the evolutionary process tests them. Indeed many animal species perform disadvantageous activities; if they do too many the species dies out. It is biologically an extreme disadvantage to be too inventive or creative, for an organic species may easily destroy itself in the course of an experiment in behavior, however intelligent and imaginative that experiment might seem to the eye of a god.

Homo sapiens displays this surplus nonadaptive formative *élan* in a high degree. Once he had discovered agriculture, his survival was ensured under the existing conditions. But his surplus vitality, expressed in a vast range of creative and inventive activities, led to entirely new modes of experience and behavior. Homo sapiens as marked by his social habits in 6000 B.C. has practically died out. Man has changed himself, and this new kind of man has multiplied and spread into all lands. This historical process cannot be ascribed to organic processes preserved by natural selection because they serve the maintenance of organic forms characterized by stable structures and patterns of behavior. In his behavior and in his brain man has created entirely new forms. And in this formative activity man displays a characteristic which is present, though in less degree, in other organic species. Animal play is not a mere by-product of the process of organic selection, it is a direct expression of an inventive or constructive tendency, a manifestation of free surplus activity expressing a tendency toward new forms.

What then distinguishes man from all other species?

Many answers have been given to this question and most of them are already known to be wrong. It was recently suggested that man is distinguished by his symbolizing faculty, which can be described as the use of signs to represent or stand for something else. But it is now clear that many animal species use highly developed systems of symbolic communication. Indeed one result of recent zoological research has been to confirm what most naturalists and keepers of pets have always known or imagined, that many animals are capable of highly subtle, intelligent, co-operative activities. Homo sapiens would be making a mistake if it claimed a monopoly of any of the following capacities: learning, educability, symbolic communication, intelligence, courage, loyalty, affection, or sense of duty.

Yet common sense suggests that there is a clear distinction between man and the most intelligent animals. No scientific view of man is of much value until we have some idea where this distinction lies. Here is one suggestion.

Man may be distinguished by a special capacity of his brain, to wit the capacity to form highly specific records (memories) of particular events, *which records can be evoked apart from the organic context in which they were formed*. It is unnecessary to discuss here whether or not this distinction is only a matter of degree, for only in man did it lead to the establishment of a developing tradition of thought. But it seems that only man can remember, imagine, and think of particular things *in detachment,* and thus use his mind in preparing long delayed responses to the environment.

All animals possess the faculty of memory in some degree. But it is possible that man alone can form highly specific memory records which can be evoked apart from the dominant emotions and operative drives of the original situation in which the memory records were formed. In so far as within the physical picture the records are preserved in the cerebral cortex and the emotional drives are mainly glandular,* it may not be an excessive simplification to suggest that man is distinguished by his ability to use his brain without his glands, that is to remember, and think, and speak about things apart from complete activity patterns of instinctive behavior.

This is an old idea: reason as the result of the emancipation

---

* Though the emotional aspects of human behavior may also be associated with the hypothalamus, a ganglion of nerve cells at the top of the brain stem.

of thought from emotion. On this view man is most characteristically human when he is exploiting this separation of
cortical from glandular processes. But when he is feeling-
thinking-imagining intuitively, in terms of a complete situation actually dominated by some emotion, he is being least
"human" and is closest to the animal world. Birds, cats, dogs,
chimpanzees, and all the higher animals may be one with
the human sculptor, musician, and other intuitives, in their
capacity for organizing total situations in terms of one
dominant emotion, or sequence of emotions.

Thus one characteristically human line of development has
been toward reliance on abstracted linguistic symbolisms from
which emotional implications have been as far as possible
eliminated. This line has been of immense power, it has created
philosophy and science, and it leads finally to logical positivism * which in its extreme exponents seems to be aiming at
the maximum of abstract form with the minimum of meaning.
It appears inevitable that the next development must be
toward the reintegration of thought and emotion in a manner
which preserves the hard-won clarity of modern logic. Abstract
analysis may have reached its limit, the next step may lie in
the clarification of the proper relation of thought to living, and
the discovery of a way of thinking which can provide man with
insight not only into external nature but also into his own
situation.

Measured in terms of elegance of living man scores low
among the organisms. This seems to be inevitable, things being
as they are. For Homo sapiens is a species endowed with an
organ for rational thought, but lacking a proper way of think-

* As an analytical and critical technique logical positivism is supreme
but it should not be regarded as a general philosophy.

ing rationally. He is not biologically complete, and is incapable of proper adaptation to his own potentialities, until he possesses an elegant system of thought which reveals the simple over-all relationships of everything without neglecting the details. To be properly himself, that is to make the appropriate use of his hereditary potentialities, man must be in possession of a unified and comprehensive theory of nature, in fact an elegant theory of physics, biology, the human brain, and of those aspects of himself which can be covered by a scientific theory.

This is a revolutionary and embarrassing thought. It implies that man is necessarily maladapted so long as he is without the insight that only an elegant and comprehensive science can provide. But that is a very awkward idea, for a scientist. Final insight, comprehensive understanding, an elegant synthesis of all the sciences, is more than most scientists are ready to contemplate. Moreover there is little sign of it yet.

Still there may be no other way, for science or for man. The next major step, in philosophy and in science, may have to be toward the recovery of simplicity and unity in knowledge. Sudden advances are a commonplace in the history of science. The novelty mainly lies in the fact that this time it is a question of a comprehensive unification which must include man in its scope.

This is certainly awkward, for it raises many unresolved problems regarding the nature and scope of scientific methods. Yet there is no reason to believe that a comprehensive intellectual insight is impossible provided that the role of the intellect in relation to action is recognized as restricted. There may be no limits to intellectual *understanding,* but the part played by understanding in influencing *action* may be limited. Reason may modify action, but reason and action are both only specialized expressions of a broader organic vitality. This view is surely an essential part of any genuinely biological conception of man.

But unfortunately during the period 1850–1950, when Western man began to recognize that the conscious intellect must be interpreted as the flower of deeper and less conscious organic processes, the most effective studies of this new realm were devoted solely to the *pathological* unconscious. As far as I know Freud never pointed this out clearly. While other thinkers from 1850 onward had been concerned with unconscious processes in general, Freud was originally mainly concerned with the unconscious in a pathological condition, and many of the technical terms of his school of thought may ultimately prove to be inappropriate to unconscious processes in general.

Thus the intellect was being expected to surrender its autonomy to a pathological libido, and to regard *all* rationality as mere rationalization! No wonder it rebelled, and even now hesitates to recognize its own specialized status in the hierarchy of the expressions of human mental activity. This hesitation will only be overcome when the operations of the intellect are seen to require, prior to and beneath their logical rearrangements, a formative process which shapes experience into units which the intellect can employ.

Like all real advances this carries a sting. For it implies what is still widely denied, that the conception of "conscious purpose" is not valid in the ultimate analysis of human behavior, perhaps being legitimate only in relation to repetitive situations. The idea of a conscious purpose, or a preconceived aim regarded as the effective determinant of action, is not in general appropriate to the interpretation of human behavior.

The more general underlying factor is an organic impulse or tendency, and it is often mistaken to project this tendency into the future and to treat behavior as consciously purposive.

An analogy from the animal realm may again be helpful. Zoologists have recently suggested that many important animal activities, such as eating, mating, and fighting, may be regarded as the expression of an inner tension seeking release in a consummatory act. This act may possess survival value, but it is the presence of the tension and its release that directly determine the act. In the case of man one might add the process of creation to these consummatory activities whose end or result is from certain points of view less important than the release of tension which they bring about.

The inventor, the creative thinker, the artist, the poet, and the mystic will understand this idea. For anyone whose methods consist in allowing something to germinate and grow inside himself knows that he has no clear purpose, other than to facilitate the processes of growth and creation, and thereby to release himself of a burden and to experience consummation. He cannot know in advance precisely what will come of the cultivation of his faculties.

All legitimate human "purposes" in some degree share this uncertainty and nonpurposiveness, and those who neglect this suffer bitterly. Parents with preconceived ideas for their children, men who spend ten years preparing for a highly specialized achievement and then discover that the situation has changed, in fact all who stay too long in one rut are disconcerted to find that when the original purpose is achieved it brings no satisfaction. They have grown older and the world has changed.

The surrender of sharply preconceived purposes is seldom

easy, nor is the discarding of the concept of purpose from the theoretical interpretation of human behavior. But these steps lose their difficulty when the underlying organic tendency and the human formative *élan* are recognized, both subjectively and scientifically.

A time may come when it will be as natural and as comforting to recognize oneself as part of organic nature and of the human adventure as it is for some today to know themselves members of a Church, a nation, or a family. This would not be a vague or sentimental realization, but a specific awareness, at once subjective-intuitive and objective-scientific, of the relation of oneself as person to the laws of the inorganic and organic realms.

That condition lies ahead, but there is a path toward it, not a road ready for heavy loads, but a series of stepping stones which can be tried out.

The steps take the form of tentative expressions of universal principles linking nature and man, appropriate to everyone of whatever color, religion, or sex. They are provisional, because they await confirmation by science. But they can be used now for comparison with other such principles and for testing against personal experience or scientific observation. Whether or not they prove scientifically valid, they possess one justification: they are discriminated from all alternatives by their elegance. They offer the simplest resolution of the basic problems involved in any interpretation of man.

The most comprehensive natural law expresses a formative tendency.

This universal tendency finds one of its expressions in organisms, which display a tendency both toward the development and extension of a specific stabilized form and toward the development of new forms.

This surplus formative vitality finds its most highly developed expression in the human individual, a unique potentially harmonious hierarchy of formative processes, partially guided by the formative processes of the organ of thought.

All human values arise from this largely unconscious formative vitality, and are valued because they express it. Personality is strengthened by the recognition of this fact and by the Yeasaying affirmation of life, without which no integrity of character is possible under the conditions of our time.

The aesthetic sense, or surprised delight in the products of formative processes, underlies all human affections, faculties, and judgments, whether these judgments are seen as aesthetic, intellectual, ethical, or practical.

It is a condition of this tendency in a universe of accident and interaction that no particular expression of it is left undisturbed to attain its terminus in a perfect and permanent form. Rhythmic deformation, lasting distortion, and final disintegration are inescapable. But this matters less when the source of all value is recognized in the *tendency toward* order and harmony.

A science of formative process alone can develop and justify this tentative picture of man.

IX.

His Plastic Brain

The brain simplifies, records, and facilitates its patterns of activity

SINCE the time of the ancient Greeks it has been evident that the mind is somehow linked with the body, and yet no advance has been made toward understanding this connection. What is the meaning of this failure?

Sir Charles Sherrington, who understood the working of the central nervous system in man as well as anyone, wrote in 1947: "The problem of *how* that liaison is effected remains unsolved; it remains where Aristotle left it more than two thousand years ago." But why?

Some have drawn the conclusion that here human intelligence must confess itself beaten. Thus Schrödinger has suggested that the parallelism of thought and brain processes "lies outside the range of natural science and very probably of human understanding altogether." That inference seems unwise if one recalls human ignorance at 1000 B.C. and all that has been understood since. Major problems often wait a long time, until their day suddenly comes. One should never seek to set limits to human understanding.

139

Yet it is reasonable to assume that a problem which is two thousand years old is insoluble *in its traditional formulation.* There may be no such problem. But there are signs that the real problem, expressed in a different manner, may be resolved in this century as the result of new experiments and ideas. The purpose of this chapter is to explain this in relatively simple language for those without specialist knowledge of the human brain.

The problem is often described as the relation of "mind" to "brain." Each of these words evokes a wide range of associations and ideas, and what we really mean by the relation of mind to brain is the relation of these two sets of ideas. I suggest that the mind-brain problem has proved insoluble because these two sets of ideas contain misleading implications. When we have improved our conceptions both of the mind and of the brain, and only then, will the problem become soluble.

In a crossword puzzle if you are trying to find two over-lapping words and have chosen both wrong it is not surprising that you cannot make them fit. The fit only comes when both words are right. Similarly the difficulties of the mind-brain problem will disappear when we have got two ideas right: how thought proceeds and how the brain works.

After all there was no reason to expect that the problem would be soluble until both sides had been tidied up. Then why does the two-thousand-year delay strike us as so surprising?

The answer lies in the fact that the seemingly innocent monosyllabic words "brain" and "mind" carry with them all the prestige of two thousand years of usage, and hence an illusory sense of clarity. But what seems clarity is only the comforting assurance that if I say, "a bullet went through his brain" or "my mind is not very clear this morning," you will know what is meant.

Thus the words are meaningful in an ordinary sense, but are not scientifically clear because we do not yet understand the laws which either brain or mind obeys. We have not identified the laws of thought or the laws of brain activity.

We can restate the problem as the task of establishing a picture of the processes of thought and of intelligent behavior in terms of the changing condition of the brain, *using improved ideas of both*. It must be possible to achieve this.

There is a fallacy that the problem arises because we speak carelessly and keep mixing "two languages," the physical or behaviorist language that uses the objective concepts of physics and physiology, and the mental or psychological language of subjective experiences, images, and ideas. For example, to say "my delight at seeing her made my heart beat twice as fast" is to combine these two languages. It is suggested that if we stopped "mixing the categories," and spoke one language at a time, all would be well.

This sounds tidy, but it neglects the fact that in their present forms both languages are inadequate and misleading, and that the best way forward is to improve both until they coalesce to form a single comprehensive language. At present we are forced to say certain things in the physical language and other things in the mental language, and until both are improved this mixing is unavoidable. We must go on speaking of "delight" and "heart beat" until we have found a single language covering the essence of both.

This is not an academic problem of no interest to the man in the street or the woman in the home. It has not preoccupied the human mind for a hundred generations without good reason. When we are well we may feel that we can afford to neglect it, but there is no illness in which the mutual relations of thought processes and physical processes does not play some role, and it is sometimes crucial. For example we speak of anxiety producing physical symptoms, because the single language is missing. In 1953 these facts are commonplace, whereas a hundred years ago they were seldom mentioned. We seem to be more subject to "psychosomatic" illnesses or more aware of them than were our ancestors.

The course of every physical illness is in some degree affected by mental factors, and of every mental illness by physical factors. The mental state affects the working of the organism, and the organic condition affects the working of the mind, because the two are aspects of one process.

This much is clear and yet medicine cannot here make any fundamental advance because we do not yet understand either side well enough. The development of psychosomatic medicine is delayed, and there is no sign yet of a unitary medicine, for this reason: the single language is missing in which alone the *complete* truth about any illness can be expressed. For only in a single language can all the relationships of health or of disease be clearly formulated.

If a problem is recognized as being very pressing, the conditions may be present which make possible a new orientation. In fact one can already suggest the line toward a solution.

The long frustration may be due to an overemphasis in both languages on permanence, and a neglect of transformation. When conceptions based on permanence are replaced by more powerful ideas representing transformations the difficulty may disappear.

Notice that the dominant conceptions of the physical language represent persisting entities: atoms, molecules, structures, tissues, organs, anatomical connections, and so on. Here attention is focused on separate relatively stable unchanging entities and their spatial relations. And on the other side the main features of the mental language represent separate and supposedly steady mental states: feeling, wishing, willing, imagining, thinking, and so on. The emphasis is here on distinct states of awareness, each assumed to be capable of persisting unchanged while it is present. The assumption is made on both sides that persisting entities or states hold the clue to understanding.

But no connection can be established between such static conceptions. There is certainly no simple correlation between molecules, cells, or tissues in the brain (regarded as static) and unchanging states of consciousness, such as feeling, willing, or thinking.

However the molecules, cells, and tissues are not static, they are perpetually undergoing pulsations and transformations. And the supposedly steady states of awareness are also misleading abstractions, for the character or quality of consciousness is always undergoing transformation. It is of the nature of perception that it turns into thought, of thought that it becomes willing, and of willing that it flows into action, the whole of this process being modified and guided by emotion. When the transformations in the brain and the transformations in consciousness are both properly understood the two pictures will coalesce into one.

In another universe governed by Platonic balance and symmetry one might be able to add "and then both viewpoints, the objective and the subjective, will be justified equally." But that desire for balance is frequently mistaken. Dialectical antitheses often conceal an asymmetry.

Brain and mind, physiologist and psychologist, external observation and introspection appear as balanced pairs but the two sides are *not* on the same footing as regards the advance of science. The development of the human intellect has rested primarily on the visual sense. We can use our eyes, check each other's reports, and eliminate errors. At so critical a moment as the present there is little doubt that the main path toward deeper understanding depends more on the improvement of the objective picture of processes in space than on analysis of the subjective qualities of experience. It will be easier to reach the unitary language through improving the physical picture than by working from the psychological side.

This emphasis on the physical-objective as against the psychological-subjective method does not justify any dogmatism on the part of brain physiologists. No adequate theory of brain function is yet in sight, and in view of the immense importance of ideas about the brain, scientists concerned with it should display the greatest caution.

For the science of brain function has only just begun, and very little is yet known about the actual working of the cerebral cortex. Until recently all the research on the central nervous system was concentrated on the nerve tracts, the single nerve fibers, and the pulses of electrical activity that pass along the cylindrical surfaces of the fibers. No experiments had been done on the processes of the cortex and nothing was known about them.

This is a striking fact, because the physiological processes of the cortex are associated with the mental processes of memory, recognition, imagination, and reason, and understanding of what goes on in the cortex is an essential part of any scientific view of man. And yet we are still completely ignorant regarding what happens after the arrival of a pulse at the cortical areas which receive sensory impressions and before the departure of the corresponding pulse from the areas controlling motor activity.

It seems that the physiologists studying the nervous system during the early decades of this century felt instinctively that the brain was too complicated and concentrated their attention on single nerves or bundles of nerves. The idea of a reflex, or combination of incoming and outgoing signals to control a response to the environment, was commonplace fifty years ago and yet nothing is known about the cortical processes by which complex incoming signals are transformed into the appropriate outgoing ones. Since 1935 studies of electrical brain rhythms have been in progress, but so far they have thrown little light on this problem. The human race, even in this scientific age, is still abysmally ignorant regarding the physiological processes on which man's distinguishing characteristics depend.

This is only partly due to the complexity of the experimental problems involved in the study of brain function. It is also the consequence of the lack of guiding ideas, and this in turn is partly due to the prevailing prejudice against speculation. Yet only informed and disciplined speculation can provide the ideas which are necessary before decisive experiments can be designed.

The most fertile new ideas are simple. Where could it be more appropriate to provide an example of this than in a perspective of the second half of this century? So here is an illustration of the kind of idea which may help those carrying out experiments on the animal or human brain. Pavlov got his ideas of the animal brain by observing a dog looking at food; I shall present an idea of the human brain by observing a man looking at a woman.

A man is walking down the street and sees ahead of him an elegant woman, a harmonious configuration of form and color in movement. He has never seen her before and yet a moment later he can recognize her in a flash though she is now further away.

The most important features both of thought and of brain processes arc illustrated in this episode. We shall first consider three aspects of it separately, though we shall later see that they are components of a single indivisible process, and we shall give these aspects abstract Latin-derived names to mark the fact that they are abstractions from a single process.

1. *Modification*. The man's cortex is *modified* by seeing her so that he can remember her at a later moment.

Somehow, somewhere, the brain material is modified by the processes occurring in it, so that it retains a record or "engram." This property is the basis of all memory and learning. Without it there could be no animal intelligence, and no human speech or culture. But there is still no accepted interpretation of these brain modifications, which also accounts for the other two aspects.

2. *Simplification*. The man does not see or remember the woman as she is, but only as a *simplified* image of the over-all features of color and form. The dominant color scheme carries him away, so that at first he does not notice that the colors of her hat and of her bag do not exactly match.

All cortical records are in some degree falsifications of the event which they seem to record, in that they start with drastic simplifications. Details, discrepancies, and distortions are blindly neglected, every act of perception simplifying the object, and every recording process in the brain simplifying the incoming signal.

This is one of the most subtle and beautiful things in organic nature: *truth grows out of error*. Indeed truth can only grow out of erroneous simplifications. It might be better to stop speaking of a property or quality called "truth," and to substitute the *truth-ward process,* the advance from greater error to less. Just as we may come to recognize beauty in the forming power rather than in the resulting static form, so we may see truth in the truth-seeking faculty rather than any particular expression of it, which must still contain error.

This advance toward truth is possible because the cerebral cortex at first accepts only the simple outlines so as to make a start, after which the details can be progressively filled in, step by step.

There is no conceivable manner in which a progressive movement toward a correct representation could take place, other than from the oversimplified to the less simple and more exact. This is a highly complex world; the brain cannot absorb everything simultaneously; it starts by making gross simplifications, using them as a working basis, and adjusting them in the light of further experience. The secret of the brain lies in the capacity of the cortex for making these fertile "mistakes." The whole of science is a system of simplifications which work well enough for the time being and can serve as a basis for improvement.

This process of simplification is undeniable; yet there is no structural theory to account for it.

3. *Facilitation.* Once the man has noticed her elegant color pattern, it is *easier* for him to see her again. The process of

recognition, or the reactivation of her image by a fresh act of perception, is a swifter and more direct process than the original perception.

The simplified cortical record somehow makes the repetition of a past process easier. A weaker stimulus or a partial one is sufficient to excite the record once it has been established, and reactivation normally tends to strengthen it. It is easier to repeat what we have already done once; we learn by repetition.

The fact of facilitation is obvious, but again there is no theory to explain it.

The failure to account for these three properties may be due to the separation of each from the other two. What actually happens is a single process: the production of a memory, that is, a simplifying modification which can be easily reactivated. Long words for a simple fact: a short cut which can be used again, a simple image awaiting evocation. The question is: *How does the cortex produce these simplified facilitating records?* What single process can account for these three properties? This is the fundamental problem for the science of the brain.

At this point the brain, that is, mine or yours, with its inveterate habit of simplifying and forming new records to ease its own processes, demands a new term to replace these three separated aspects and to describe the single fact. It should not be a Latin abstraction, but something more concrete and plastic. There we are: *plastic!* In the context of structural biology, and particularly of brain theory, we define "plastic" to mean: "possessing the property of forming simplified semi-permanent modifications which ease the repetition of the process which formed them."

"Plastic" seems a good word for this.* It comes from the Greek *plassein*, to mold or form, and anything stable which is molded is necessarily a record of a molding process. Moreover in earlier times the term "plastic" used to convey the suggestion of an aesthetic activity, and this is appropriate since the essence of the aesthetic sense lies in the recognition of a simple order. Our question can now be restated: *What explanation can be given of the plastic power of the cortex?*

This should be the first requirement of any theory or model of the human brain. Let us see how some current models stand up to this test.

The brain is a telephone exchange. This is a very superficial analogy which does not throw any light on the plastic activity.

The brain is a clean sheet on which records are taken. This explains nothing. Moreover even at birth the brain is not a clean sheet: some brain patterns and modes of behavior are innate or hereditarily determined.

The brain is an electrical computing machine, which can record and rearrange its own activities and control its own processes in accordance with the results of past activities.

Some mathematicians, physicists, and physiologists consider that this is a nearly perfect model, in the sense that a sufficiently complex machine of this kind could simulate every kind of human behavior, and even construct duplicates of itself. But the passion with which this view is held betrays a hidden psychological motive.

On the other hand most biologists tend to emphasize the differences between any machine constructed by man and organic systems resulting from a long evolution and a process

---

* No confusion need arise with other uses of the term. Probably only the brain is actively plastic, in the sense defined above.

of growth. Every organism stands in a closer relation to its own past history and present environment than does any constructed machine. Moreover man provides the machine with its aims, whereas he discovers his own.

The contemporary controversy between these two attitudes may look very strange in twenty years' time. For it is impossible to decide how similar and how different two types of systems are, like the brain and the computer, until you know the structure and laws of both. But we do not yet know how the brain works. I have little doubt that when we do fundamental differences will appear between the processes of the cerebral cortex and those of electronic computers. Indeed this is mere common sense, for the two systems are different in very many respects.* But this does not mean that the analogy is not of great value in provoking discussion and suggesting experiments.

A satisfactory model of the brain should throw light on the biological principles and functions inherent in its activity. Here we have stressed the plastic property of the cortex as being of basic importance in understanding its operations. The electrical computer can simplify, record, and facilitate its own patterns of activity, and can therefore be regarded as plastic. But this property is arbitrarily put into it by man and is not an expression of general biological principles.

Perhaps in addition to *mechanical models* one should look for an *organic prototype* of the brain.

This is biologically appropriate. For every multicellular organism developed by cell division from a single undiffer-

---

* One difference is that an electrical computer must contain highly insulated channels (wires, etc.), whereas in a living organism no part is isolated from the influence of neighboring parts.

entiated cell, and this parent cell can serve as a prototype for the adult organism. All the different organs and functions of the mature organism can be found "in embryo" within the parent cell.

In animals, for example, special synthetic regions in the original cell can serve as models of the glands or specialized synthetic organs of the adult, the nucleus of the parent cell corresponding to the reproductive glands. The elastic framework of the original cell, which controls its shape, models the muscles of the adult, and the polarized external membrane models the conducting surfaces of the nerves. These relatively stabilized parts of the original cell provide prototypes of and develop into the structurally stabilized organs of the adult organism.

But what part of the parent cell models the cortex, which is a semifluid gel? The natural answer is the relatively unorganized semifluid conducting material of the inside of the cell, called the cytoplasm. This would imply that the functionally important part of the cortex is not the sharply differentiated parts of individual cells, but the mass of cytoplasm which extends throughout the cortex. This nearly continuous mass of unorganized semifluid cytoplasm must somehow possess a plastic property of simplifying, recording, and easing the patterns of electrical activity which pass through it.

But how?

Let us consider some quantitative facts about the cerebral cortex. It is the product of about one billion years of organic evolution, and it reached its present size in some species of *homo* around 200,000 years ago. It contains some twelve billion cells. The nerve fibers which conduct impulses to and from the cortex are about 1/1000 inch across, and undergo changes of electrical potential of about 1/10 volt, which implies an exceedingly high electric force across the thin nerve membrane. Pulsations may follow one another at ten a second or more. A new memory trace can be stably fixed somewhere in the cortex in a few seconds, and an unexpected association may require several seconds or more to evoke.

But these numerical facts mainly about individual nerve cells do not directly help us to understand how the brain works. For though the incoming and outgoing signals travel along single nerve fibers, the higher functions of the cortex do not appear to involve particular cells or groups of cells so much as patterns of excitation extended over regions, patterns which follow directions and undergo transformations that are not precisely localized in terms of the cellular structure of the cortex. The normal working of the brain probably involves the synchronous pulsation of regions of cytoplasm extended through an immense number of cells. What matters is not the fine cell structure but the laws of transformation of the activity patterns in extended masses of neural material. The crucial question for a theory of thought is: How are these extended activity patterns unified, recorded, and facilitated? How does the semifluid cytoplasmic mass of the brain protein come to possess a plastic property, in the sense used here?

The answer may be simple.

If you go to a pile of junk at the back of a garage, you may find something you want, but you cannot pull it out because it is caught somewhere. So you work it to and fro and as you do so the pile gradually falls into place, easing the movement. Disordered systems which are subjected to cycles of change often tend to order themselves so that the cycles become easier. That may be how the brain works. The brain may be plastic because its disordered parts arrange themselves to conform with the processes to which they are subjected, the brain as a whole thus ordering itself in accordance with its experience.

In the cortex there is a considerable amount of relatively unorganized protein, in which the molecular chains are probably arranged in a haphazard manner. When an activity pattern (electrical polarization pulse) passes through such a region in a particular direction it may tend to draw the molecular chains into more regular orientations so that they can respond more easily to a repetition of the same pulse. Thus the pulse may exercise an ordering influence, and leave behind it a modification of the original system which eases the repetition of the pulse. Moreover this modification will correspond only to the statistically dominant or simplest over-all features of the pulse neglecting its details, so that it constitutes a simplified record of the pulse, facilitating its repetition in this simplified form.

If many similar pulses are repeated the protein will gradually work itself into stable well-ordered regions corresponding to the simplest and most frequently repeated features of the pulses. The plastic property of the brain is thus explained in terms of a self-ordering tendency of disordered protein subjected to pulsations. On this view the unconscious processes of the human imagination which order experience into persisting symbols capable of re-evoking the same experience are identical with the plastic processes of the self-ordering protein of the cortex. The plastic transformations of brain and of the mind are identical.

In 1953 these suggestions lack observational confirmation. There is as yet no experimental evidence of this self-ordering tendency in the brain cytoplasm when subjected to pulsations. But no one has looked for it.

How serious are we in claiming that our understanding of ourselves is important? Does the understanding of the brain really matter? If so, let us get down to it. Too little attention is being given to *new ideas* about the brain.

In 1921 the *Scientific American* gave a $5,000 prize for the best short essay on relativity. Why should such prizes be given only for studies of problems that have already been solved? Why not one for the best essay on "The Theory of Brain Function and the Experiments to Test It"?

X.

His Creative Activities

All man's constructive, inventive, and creative activities arise from the plastic activity of his brain *

THE argument of the previous chapter leads to this intriguing conclusion: the coming decades may see the creative powers of the human mind traced to the structural properties of the material fabric of the brain. This would imply that "creativity" had been identified in "matter"!

Since creative processes and material processes have usually been regarded as essentially opposed, it is difficult to exaggerate the consequences of such a discovery. It might lead to the collapse of the separate mental and material languages, and their coalescence into one that did not imply two distinct modes of being.

First of all the implications of the term "creative" would be transformed. At the present time one might say that it stood for what is at once the noblest and the emptiest conception in contemporary thought. Nothing in man's experience ranks

---

* But this tentative formulation does not imply a "materialistic interpretation of mind" in the traditional sense, as the text shows.

higher than his creative imagination. And yet the conception often seems sentimental, hollow, vague, and mysterious, like the pseudo-name of a god so awe-inspiring that his true name must never be mentioned.

Why should we not view the creative process face to face? Is the creative principle so timid or frail that we hesitate to uncover it in its flesh and blood and detailed structure? Can anything be so noble that we refuse to admit it to a place in the universe of structure?

If the advance of science leads the human mind to recognize its own creative faculty as the expression of structural properties of the brain, then it may be necessary, for certain purposes at least, to discard the term "creative" with its misleading dualistic associations and to use instead the more general term "formative," in reference to the structured processes which create new forms. This would be one step toward the single language.

Then secondly, the implications of the term "material" would also be transformed. It stands now for a highly objective but unduly restricted conception. The properties of material systems are the things we can all agree about most easily, indeed "matter" is almost a name for what is objective, law-abiding, and reliable. And yet it implies a degree of inertia or unchanging persistence that is irrelevant to many natural systems.

When "matter" is found to be "creative" in the plastic cortex, the usefulness of the term "matter" will have been exhausted and it will then be better to drop such phrases as "material systems" and to speak only of "physical structures" or "changing spatial structures." The paradox resulting from the discovery of *creative matter* can be resolved by renaming it *formative structure*.

What we call the "creative imagination" is, on this view, the entry into awareness, the coming to attention, of the products of the plastic activity of the cortex. When one asks "where do those bright ideas come from?" the answer is: from the unconscious plastic processes of someone's brain. At least this is the best working hypothesis.

Until recently there has been no serious attempt by the leading schools of psychology and physiology in the English-speaking world to come to grips with the problem of the nature of the creative imagination and the processes underlying the creative activity of the human mind or brain. To the exact scientist and to the psychologist of the period 1910–1930 seeking to create an exact science the imagination was almost taboo, as though it were too unreal, mysterious, or disturbing to be mentioned in serious scientific discussion. Thus no work by an academic psychologist or physiologist of repute appears to have recognized the scientific importance of the creative aspects of thinking, above all in man, prior to Wertheimer's *Productive Thinking* (Harper, 1945). The leading psychologists of the schools influenced by quantitative science seem to have unconsciously turned a blind eye to this important problem for an entire century, from 1850 to 1950.

In retrospect it is not hard to discover reasons for this. Probably the most important was that exact quantitative methods tend to rely on analysis, precision, and permanence, whereas the creative process is a combining, simplifying, and novelty-producing activity. Thus the existence of the human creative imagination as a fact of nature like gravitation or chemical affinity challenged the adequacy of the exact scientists' proven methods. In these circumstances it is not surprising that they tended to develop a blind spot. There is no blame on them for that. The neglect of more difficult problems was necessary so that the new science of psychology might establish itself on simpler ones.

A metaphysical revolution is necessary to extend the traditional foundations of science: analysis, precision, and constancy, to include ideas which can do justice to combination, simplification, and a process leading toward novelty. The new doctrine, if it is to represent a genuine scientific advance, must retain all the benefits of the old, but go further and yield new fruits.

It seems that this revolution is already in process. No one who has experienced the development of scientific thought since, say, 1910 can fail to recognize the increasing prevalence of structural and developmental ideas. It is as though the collective mind of the exact scientists were trying to learn how to combine atomic entities into patterns, how to simplify complex situations, and how to describe processes which lead forward toward a unique end state. The conclusion appears to be inescapable that the intellectual context of the 1950's is much more favorable to an exact theory of productive thinking than it ever has been before.

At least this appears to be the unconscious inference of many professional psychologists. For during the last decade academic psychologists have begun to tackle with the earnestness it deserves the problem of a *theory of thought*. What is the essential character of the thinking process in the most general sense, and how does the cortex operate when thought is proceeding?

There have been many guesses at the essential character of

thought, but most of these are far too narrow and exclude many aspects of what is properly regarded as thinking. Thus the process of thinking has been variously interpreted as *following mental associations,* a very restricted notion which neglects all the features that have here been stressed; *solving problems,* but what about identifying the problems? *filling gaps,* but the gaps have to be recognized first; and so on.

None of these suggestions throws light on the character of thought in general, for they all either neglect or underestimate the importance of the constructive, inventive, and creative aspect. To put this right it may be necessary to go behind the more explicit and conscious aspects of thought to the plastic activity of the brain, which has been forgotten because we are less aware of it and because its properties are alien to traditional scientific methods.

There is today a great concentration of physiological and psychological research gradually approaching this problem, and the coming decades will certainly see the development of scientific theories of the imagination, which may or may not confirm the suggestions made here. The greatest need of all is again for *improved ideas.*

But in the meantime let us see what light the conception of a plastic activity of the cortex can throw on the aesthetic creative process.

It suggests, for example, that the essence of creative activity does not lie in a mere *selection* of material from already given elements, but in a simplifying process which automatically involves not only the selection and rearrangement of the available material, but its modification in process of developing a simpler form. A work of art need not be simple in any absolute sense, but it is always simpler than a random collec-

tion of similar material. The creative process, in simplifying
the material used, distorts it, as the human brain does in all
its other mental activities. Not merely modern art, but all art,
and not merely art, but all thought distorts the phenomena
which it claims to represent, this distortion being implicit in
the simplifying processes of the brain. And without this simplifying distortion the mental images would not facilitate
insight into the experience or situation which they represent.

Thus all creative activities reflect the properties of the
underlying plastic activity. Moreover this creative aspect is not
restricted to the special achievements of genius, it is present
wherever a man or a woman or a child is thinking: in the
infant as well as in the adult, and throughout human history
from our prehuman ancestors to those alive today. The unconscious or scarcely conscious processes of the growth of
thought which form the background of the developing mental
activity of the infant and prepare it for adult life are closely
similar to the scarcely conscious preparatory phases of mental
growth preceding the creative process in specially gifted individuals. Everything that uniquely marks the destiny of man,
from infant to adult and from primitive man through existing
communities to those of the future, is the expression of the
plasticity of his cortex (coupled with his erect posture).

This suggestion need not shock anyone. To the scientifically
trained mind it needs no argument. And the religious mind
lacking scientific knowledge will grant that if God created the
universe he must also directly or indirectly have created the
human cortex. Moreover fears of an oppressive scientific
determinism threatening the freedom of the human spirit
become empty if exact science discovers a formative principle
at work in the structured processes of the brain. In the era of

formative structure which we may soon be entering the old problems of matter-versus-spirit and determinism-versus-free-will lose their old significance.

An important element in this science of formative processes will be the history of the human mind and its progressive development, viewed through the chief formative factor: the plastic processes of the cortex. Today we are compelled to speak of ritual, art, religion, philosophy, and science as if they were independent activities which could be understood in separation. Yet this is wrong. They are differentiated expressions or special aspects of one underlying activity: the cultural development of the species through the exploitation of capacities latent in its hereditary equipment, and none of them can be properly understood until they are seen in their relationships to one another, as results of one basic activity.

This grand task: the interpretation of human prehistory and history in a long biological vista, has hardly yet been begun. It lies in the common ground of human biology, philosophical anthropology, and social history, but no one has yet devoted a lifetime to it and that is what it will need.

One of the most interesting preliminary attempts to trace a single principle through all the different forms of cultural activity, from ritual to logic, is that presented in Susanne Langer's *Philosophy in a New Key*. She shows how many aspects of culture can be interpreted as the result of a symbolizing activity inherent in all mental processes, carrying further the ideas of Cassirer and Whitehead on the role of symbolism. The present arguments are complementary to those of Langer. For the plastic activity of the cortex is what is needed to give a biological and physiological basis to Langer's symbolizing activity of the mind.

Perhaps some able young biologist with a balanced view of the relations of the human and organic realms will be seized by the ambition to tell this story of the development of man, bringing together into one vivid presentation all the understanding of the various sciences and branches of scholarship concerned with man. The biology of man is a young study that has not yet found its voice, and the opportunity is waiting for the man.

Here one can only hint at some of the fascinating themes which will require treatment. For convenience these suggestions will be divided into stages.

1. *Before man.* In the primitive Hominidae and types of Homo prior to Homo sapiens, which were in existence around one million years ago the cortex must have already possessed a high degree of plasticity displayed in an advanced animal intelligence. The normal operations of the brain were carried on under the dominance of some immediate vital need, but a surplus of plastic activity was making possible the development of symbolic rituals, for example, of dancing, shouting, and play, independently of instinctive needs.

At this stage any individual Homo might discover that he could *play* with organic signals, and make a cry of love, or attack, or hunger, or pain, or practice the call to the hunt, when these situations were not actually present. Yet this "discovery" was scarcely conscious. The individual would start the new kind of action spontaneously, his nonadaptive surplus vitality overflowing into a new pattern of activity, or into an old pattern detached from its original organic context and so acquiring a changed structure and a new significance.

But his comrades would take his action to mean that the organic context was actually present, for example following him at the run if he called them to the hunt, and when they found that he had fooled them would be liable to strike him down in angry rage. He was the ancestor of all pioneers and creators who insult the comfortable life in old ruts, and like them he would suffer for it.

From the narrow strictly adaptive point of view, measured by the survival interest of existing modes, nothing could be more inappropriate and damaging than to use organic signals apart from the corresponding organic situation. The immediate effect would be directly contrary to the survival interest of the species, for it would tend to upset established procedures of high adaptive value. This fact is reflected in the feeling one has that such behavior is immoral, almost a kind of self-abuse, the perversion of finding satisfaction in an organic function divorced from its natural purpose—which means its original purpose.

Yet from these nonadaptive surplus experiments there ultimately developed, as a kind of symbolic play, the entire cultural tradition of man. UNESCO might set up a monument to the Unknown Homo who died on account of his originality. His example made Socrates possible, and he shared the same fate.

2. *Gradual emergence of Homo sapiens* (? from four hundred to one hundred thousand years ago). In the sub-human realm nonadaptive formative play constitutes a relatively minor element in the whole pattern of behavior, perhaps because the formative activities were soon exhausted. But the evolutionary process produced changes in the cortex which transformed this trivial surplus play of the higher animals into the progressive cultural adventure of Homo sapiens.

The major steps in organic evolution, for example from

reptile to bird, are still little understood, and the transition from the anthropoid apes through the Hominidae to Homo and Homo sapiens may in the future present interesting problems for evolutionary theory. But according to the prevailing Neo-Darwinian view this was mainly the result of natural selection operating on random mutations * or genic rearrangements. The historical fact is that a species was emerging prior to one hundred thousand years ago which later proved to possess three connected properties: first, the ability to survive longer than other types of Hominidae and Homo; second, an erect posture; and third, a hereditary equipment resulting in an adult brain much more plastic, more capable of shaping highly specific records and using them in detachment, than the brain of any earlier type. So highly plastic that many thousands of generations, occupied in accumulating the products of its activity in the form of a progressive tradition handed on from generation to generation, has not exhausted its potentiality for further novelty.

As we have seen it is too early for science to express a judgment on the extent to which this third property aids or prejudices survival in the long run. Biological common sense suggests that even if technical advances raise the standard of living and increase the expectation of life, or have done so till now, one can have too much of a good thing. For the purposes of survival stability is more valuable than continual revolutionary and disturbing transformations. Homo sapiens (6000 B.C.) might well say to Homo instabilis (A.D. twentieth century)—though we need not agree with him—"I could have

---

* The unknown "Blackrow factor" referred to in Chapter III may emerge when the meaning of "random" is examined rigorously in a structural context.

survived if I hadn't been so inquisitive. Take a lesson from my fate. I have died out, but at least I handed on to you some useful stable habits, agriculture for example. Are you handing on anything that is stable? Have you not gone too far with your experiments and prejudiced the entire human tradition?"

The formative and creative activities which are so marked in man cannot be regarded as the product of an evolutionary process which preserves only those characteristics which have already proved their adaptive value. These activities are better understood as the expression of a universal tendency present in organisms, which tends not only to maintain them as they are but also to develop new forms.

3. *Primitive man.* The Homo who made daring experiments, not from any choice of his own but in following the plastic drive of his cortex, was (presumably) isolated and despised and his example rejected. One might say that Homo sapiens had been established as a cultural species when the unique individual who displayed an unusually great faculty for symbolic play was honored as well as feared. For the human creative player, who could use his brain freely out of immediate context, gradually became to be regarded with ambivalent awe, as magic man, priest, prophet, artist, thinker, and inventor. And these created humanity.

From the gestures and cries of the solitary Homo practicing an organic function out of context and thereby transforming its significance there developed the magical drawings of the hunt made in the security of the cave, and the arts of song and speech. Speech is perhaps the richest of all human activities for it steals the delights of immediate experience and makes them into a kind of emotion-haunted verbal play. And from speech there developed on the one hand script, philosophy, science, and mathematics, and on the other the word-bound ecstasies of the poet and the mystic's attempt to convey what words cannot tell.

The plastic pulsations of the brain fabric, simplifying, stabilizing, and extending their own patterns, have been ceaselessly at work in nearly all the million million human beings who have lived since, say, 30,000 B.C. In every healthy individual this formative process has been active, in childhood guided by parental example and training and in adult life molding its own path. The collective result of these processes has been the building up of a progressive tradition of symbols, each of which is normally simpler than the thing for which it stands. But the symbols must not be separated from the activity which accompanies them. Thus the result of all the plastic processes in human brains is better regarded as a tradition of symbolic activities, for no record in the brain and no image retains its significance if it is dissociated from the total pattern of organic activity. The human story must be seen as one great pattern of activity in continual transformation.

4. *Appearance of special cultural activities.* We are so accustomed to separating various aspects of human symbolic activities that we tend to think that religion, art, science, mathematics, technology, administration, and so on are the real or basic activities and that primitive life was confused because it combined all these into one semimagical unity.

But perhaps it is we who are at fault in imagining that all the complex differentiated activities of contemporary life could ever have developed, or could coexist in one community, unless they were expressions of one underlying mode or tendency in man. And since all elements of culture depend on simplified memory records which ease repetition, we can provisionally assume that all cultural activities express the plastic activity of the brain.

Thus we should not say that "primitive life combined aesthetic, religious, and practical activities in one pattern," thereby implying that these categories are fundamental. Instead we should say: "In primitive communities the basic plastic activity had not yet developed all the differentiated but interdependent expressions which we can recognize and partially separate in contemporary life."

The difference is important. For the second way of thinking implies that if we use these differentiated categories then we must correct their excessive separation by remembering that every specialized activity contains something of every other: all religion is in some degree aesthetic, linguistic, rational, and practical; all science is in some respects aesthetic, religious, etc.; and that no sane activity can entirely rid itself of any of the basic qualities of the formative process, because that is the source from which they all derive.

No sane activity. The restriction is necessary. For while wholly nonrational mysticism, wholly abstract logic, entirely unempirical thinking, entirely nonrepresentative art, and so on are conceivable as extreme cases, they are potentially psychotic and lead to disaster unless at some stage brought back to their healthy organic root: the plastic activity of a brain within an organism coping with an environment which provides ample challenges.

This reinforces the view that the imaginative activities of the human mind are not the mere by-products of evolutionary selection. For they often threaten both the existing tolerably well-adapted community mode of life and the stability of the individual who exploits them. From this point of view there is no reason to doubt some degree of association between genius, or high creative capacity, and mental abnormality, or fargoing failure to accept existing modes of life. However this association is not due so much to a failure of genius to adapt to established modes, for he is not trying to do that, as to his preoccupation with formative processes.

In one sense the struggle with established ways remains as hard as ever. But in this century no creative person who has the humility to seek an objective understanding of his own situation need feel alone. For he can realize that he is one of a great line from the playful Homo to Max Planck whose experience led him to suggest that new scientific ideas conquer only by the death of their opponents.

We are here in a realm beyond good and evil. Worshipers of progress sometimes suggest that all creative activity is intrinsically good, but that would surely imply that the old ways which it seeks to destroy are necessarily evil. They are balanced by those conservatives who worship the past and regard all novelty as harmful. Terms such as good and bad, which imply absolute standards, do not apply to the tension between the old and the new.

Development arises from a tension which is experienced as a painful conflict of two principles, stability and change, and is often interpreted as a conflict of incompatible values. But there is no absolute criterion which can be used to balance the values and determine a correct attitude which is valid for everyone.

The yearning for one universally valid standard of values still haunts mankind, yet the conception is philosophically unreasonable, biologically inappropriate, and practically unrealizable. There is no supreme authority, recognized by all mankind, to formulate such universal values, and no values can be appropriate to all since individuals and communities vary greatly. What is appropriate to and is acknowledged by some is intolerable to others. Some persons and groups are vital, developing, and creative; others cautious and conservative.

The creative must create, and the conservatives must provide the indispensable element of stability. Without his creative *élan* man could never have become man, but with too much he cannot survive.

XI.

His Present Wrongheadedness

" 'Tis all in peeces, all cohaerence gone!"
— DONNE

THE ancient Hittites left records of a Myth of the Missing God, in which the disappearance of the god of fertility brought about a paralysis of all life that was only overcome when the god had been found and brought home again.

This tells an experience that is frequent both in the personal life and in the course of history. There comes a time when we feel we have lost our roots, the virtue has gone from us, life seems empty and falls to pieces. We cannot imagine how anything can ever be right again. Then if we are fortunate life returns, the paralysis passes, and everything is normal once more. But a generation which has experienced disintegration too deeply may be beyond hope.

In the line quoted above John Donne expressed his sense of the collapse of the hierarchical scholastic world of the Middle Ages. In our own time there may be equal ground for distress. The art and literature of this century suggest what many feel in their hearts: that man has lost faith in everything.*

* ". . . What appears to be one of the distinctive symptoms of modern
170

When Nietzsche cried, "God is dead!" he foretold our ex-perience. A tragic religious conviction based on transcendental principles may still survive, but as regards human life on this planet the years since 1914 have displayed the failure of the Western tradition to provide an inspiration and a way of living that can serve the race as a whole. The significance of life has diminished, and the social coherence and personal integrity inherited from earlier generations are threatened. At least that is how it seems to many.

Theoretically we know that growth and decay go together. But it may be beyond the capacity of any man to experience both at once. No contemporary writer shows both a penetrat-ing awareness of the exhaustion of traditional forms and a clear sense of a way ahead. Many have the first, perhaps no one as yet has the second.

But why such concern, you may say. These are the birth pangs of a new world, maybe the adolescence of mankind. The race always muddles through somehow. One civilization declines, another takes over. No human being has fully under-stood this process, but it goes on none the less. Do not try to take on what is far beyond the powers of any individual.

This may seem to be healthy advice, and yet it may be necessary to repudiate it.

For no analogies from the past are valid. There is a novel situation which may call for unusual measures. There have been progressive changes in the condition of man. The race is more aware, of nature and itself, than ever before. Moreover there are no communities unaffected by our decay which could take over. We are all on the spot together, and find it

---

literature and thought: the consciousness of life's increasing deprecia-tion." Erich Heller, *The Disinterested Mind* (British Book Service, 1953).

hard to evade the embarrassing question: By what convictions shall two thousand five hundred million live? Wherein shall mankind place its trust?

No prophet, sage, or tyrant has ever before had to pose such a question. No man and no group of men can possibly answer it in a way that will satisfy the greater part of mankind.

Yet it will be answered in fifty years, somehow. If too many conflicting principles are at work in human minds, there will be no ease or joy in living. But if a reasonably adequate consensus of conviction on certain basic matters emerges, there can be a recovery of initiative and hope.

We can make the question more precise. The men who are going to rule the world in A.D. 2000 will be receiving their education in 1975, only twenty years hence. Will the upbringing of the young in the leading countries at that date be *more* or *less* Christian, more or less Marxist, more or less scientific than now? If the race is to discover a common ground, how can it best be reached? In the ultimate analysis there may be as many different paths forward as there are individuals alive, but here we are considering the changing social tradition. How can a universal human foundation best develop beneath or within the present conflicting traditions, so that their conflict may be reduced?

Before so momentous a question all prejudices should be dropped. We should not judge the future by the past, but open our minds to what may be possible.

If we measure the world religions and the present condition of science not by their own standards but against the needs of man at mid-twentieth century, it is evident that both have failed. The religions have not provided the race with a common inspiration and science has not established a unified picture of nature, providing insight into the human situation. By these criteria there is little to choose between them.

But there seems to be more universality, receptivity, and ability to overcome past limitations in the scientific tradition than anywhere else. Here mankind has achieved nothing less than a miracle: an orthodoxy which develops itself! There is hope in this. Truly religious minds will recognize this fact, and why not? For if a God made man, He made science also. A living science may be more divine than a dead religion.

One might be tempted to try a short cut and to say: "Forget the words 'science' and 'religion.' We are in an epoch which requires new terms. This is a time of single comprehensive doctrines expressing the over-all attitude of new types of men, like Marxist man. What we need is not science or religion, but a new all-in wisdom."

There is enough truth in this view for it to be dangerous. Mankind does need an all-in attitude, and many such slippery wisdoms will be offered by the would-be manipulators of the human mind in the coming decades. 'Osophies, pseudosciences, and psycho-religions will be endemic. But they will lack the sanity, austerity, self-discipline, and guarantee of progressively deepening understanding which is the unique privilege of the authentic scientific tradition. Now if ever science must hold to its cautious step-by-step advance under the ruthless fire of informed criticism.

The scientific route toward a deeper understanding of man will be difficult. Fashions will delay progress, advances will lie unnoticed, and the new power of the scientist will continue to tempt him from his path. But he is faced by an opportunity and responsibility beside which the mission of the early Christians was parochial: the long prepared, keenly awaited, and now desperately needed enlightenment of the race on the physical, biological, and psychological aspects of man. There can be no doubt about it: this is the adolescence of man; he will shortly become much more fully aware of himself and his own faculties.

This prospect is clouded by the prevalence of two contradictory attitudes, both of which involve serious misinterpretations of the present situation.

The first is that science is descriptive and not normative, that since it only describes facts it cannot have ethical implications. Science is ethically neutral. This implies that scientific methods cannot make any contribution to moral problems.

The second takes for granted that a science of man is not merely possible, but is urgently needed to take over the responsibility of shaping the future of mankind.

These attitudes involve complementary errors, and taken together merely obscure the real position. The first is based on the mistaken assumption that there exists a cultural entity called "science" which remains unchanged though itself transforming the human situation. Yet once science begins to study man, man is himself thereby changed, and the moral problems that interest him are transformed. For example, the scientific conception of the human person as an organism potentially capable of achieving organic harmony may prove to contain far-reaching ethical suggestions. In its techniques science may be ethically neutral, but its consequences transform every aspect of human experience. It is of great importance that those who consider that science ought to remain neutral should open their eyes to the new ethical problems which the advance of the human sciences is now opening up.

But the second attitude is the more dangerous, for there is

little doubt that the coming decades will see the progressive advance of a comprehensive scientific doctrine regarding man, whether or not it is called a "science of man." This prospect is an inevitable expression of human vitality in the present context; it would be as silly to condemn it as to criticize a mountain. Within fifty years there will be a science of man. We are not free to question that, but we must ask what we are going to do with it. Is the science of man to determine his future?

There is, I believe, a clear answer, however difficult it is to apply in practice: scientific authority can provide the best picture of the facts, but the implications for human action must be discovered by every person and community for themselves. For the proper line of action for any human individual or group depends on their own condition and degree of awareness and *this can only be determined by the way in which they use their own judgment in making actual decisions*. It would be morally intolerable and biologically mistaken for scientific authority to step outside its competence and to attempt to dictate to man on the fundamentals of policy. No more unbiological idea could be conceived. Science can clarify situations, but human judgment remains the arbiter of action.

When the popular cry arises, "Where are we going, and how shall we get there?" for example when the press demands social advice from scientists on fundamental matters, they should answer: "As scientists we hope to bring enlightenment, not final or authoritative instructions on general issues. Every man, woman, and group must form their own judgments and find their own way. Our authority as scientists is restricted to revealing, as lies in our power, the interrelatedness of things, the story of the human past, and the facts of the present. On the basic principles of policy appropriate to any particular community science as such is silent, and has no authority. Science enlightens, but man decides."

This throws the responsibility for judgment and action back on to the individual. But today he often feels unable to carry it, for subjectively he has lost faith in his own powers and objectively he finds his pessimism confirmed by a lack of imaginative judgment in the contemporary mind when it attempts to deal with human problems.

It may be too early to evaluate the years that lie very close, but if we take the period 1910–1940 many will probably agree that the principal new achievements lie in the realms of logic, mathematics, physics, and technology, and that the human imagination has been *relatively* unproductive in the sciences of life: biology, psychology, and sociology, and in literature and the arts. There have been few, if any, major new ideas in the life sciences,* and only relatively minor achievements in the humanities. Literature and art have reflected man's unhappy condition and brought it more fully to awareness, but no suggestion of a future mode of experience has yet been expressed, or if it has been then it has not been recognized. There is no *avant-garde* movement in the arts and humanities leading human awareness forward, and there cannot be until a new enthusiasm returns. The creative spirit of man is only fully alive today in the abstract realms where his doubts of his personal worth do not paralyze his vision.

Yet a phase of paralysis does not last forever, and recovery is sometimes sudden. A condition of inaction and of self-hatred

* Such as, for example, the discovery of the nature of biological organization, the way the genes control development, or the manner in which the brain works.

allows no scope for the underlying vitality. Sooner or later, in the individual or in another generation, the increasing pressure of the frustrated tendencies accumulates until the inhibitions, hitherto self-enhancing, are washed away in a surge of released vitality. In the individual this reversal of condition is nonrational; arguments that seemed to justify pessimism appear irrelevant after the crisis of recovery. But recovery in the community comes mainly through the appearance of another generation with a new awareness, the rest being left to die out.

This is a ruthless process, unmistakable even in science, where the new and more comprehensive doctrine should always be welcomed. How much more so in the social realm where no lip service need be paid to new attitudes! Under present conditions any substantial extension of the age limit for professional life would gravely impede not only scientific progress but the general development of the human mind. For when the missing god is found and brought back in a new guise by a triumphant younger generation, the majority of the elders in council fail to notice that anything important has happened.

No human community can survive for long whose individuals secretly repudiate human life and personality as they know it. Yet this seems to be the inner condition of many in the Western world.

In this spiritually arid century where can a symbol be found that can stand up both to scientific realism and to our intense personal awareness of human frailty? It would be a mistake to think that Western civilization can survive without such symbols. The dilemma of our time lies in the question: What symbol of fertility or productivity can enjoy unquestioned validity and help to restore man's belief in himself?

Can we find an answer?

Libido? This is both real and fertile. But it can scarcely be said to be missing from contemporary thought. Art, literature, and advertisement evidence an unusual degree of conscious preoccupation with thoughts of sex. Moreover Freudian libido is one aspect only of human fertility and is conceived as being in essential conflict with the rational mind. Freudian man is in pieces. The symbol which is needed must represent a more comprehensive principle of which libido is only one expression.

Growth? Or Life itself? Albert Schweitzer's reverence for life is a contemporary recognition of organic fertility as a sacred value. But something is lacking. The return of the god must produce the shock of delighted surprise at a new vista, so that we take a deep breath and start living anew.

In such a situation as this it is probable that no single mind can see more than a tiny part of the picture. But it seems to me that one clue lies in an accent on form. Our children, or theirs, may enjoy the experience of watching exact science discovering the missing symbol in a principle of the development of form.

If one is hoping for a particular event one is certain to see signs of it everywhere, and this impression may be misleading. Yet there is objective evidence that this is a phase of synthetic endeavor in science, and that pattern, structure, and form are in the air. A major simplification may be in preparation, and its path may be eased by disciplined anticipation.

One conclusion seems unavoidable: that the central element in any unification of knowledge and in any restoration of the continuity of the human vital impulse with rational thought must lie in the intellectual and personal recognition of a universal tendency toward form. This essay is one element in a continuing attempt to develop that idea and clarify its implications.

Thus behind animal libido and underlying organic growth we can recognize the *formative character of natural process*. Once this has been identified by science man's conception of himself will be transformed, for then he will be able to appreciate the formative power of the unconscious organic processes in his own person. These unconscious processes fall into two groups: the ordinary physiological processes which maintain and extend the stabilized structures characteristic of the species, and the cortical processes which develop new patterns of activity and thought.

It is a strange illustration of the distorting power of a great idea that Freud, in developing his vision of the pathological unconscious, should have scarcely noticed the formative, synthesizing, and integrating power of unconscious mental processes, for example in all kinds of creative and constructive thought. Unhesitating acknowledgment of Freud's genius and of the importance of his work for human welfare need not prejudice the recognition of its restricted validity. The unconscious processes of the human mind are not in essential conflict with rational judgment, for in the adapted individual all human faculties are components of an integrated hierarchy and derive their impetus from one source, the formative tendencies of nature as manifested in the human person. The human race awaits a great psychologist who can see all this, holding the whole and the parts in balance.

Our world is in pieces. There is no international or social harmony; no balance between past tradition, present experience, and constructive action; no unity of knowledge; no adequate insight into ourselves. In particular there is no elegant theory of medicine and psychology, to provide understanding of disease. These lacks are not new but they intimidate creative endeavor more than ever previously, for we are more aware of them.

There are many who can no longer expect the needed relief from a religious prophet, a vague spiritual renaissance, new political leadership, or any irrational transformation. But there is one event which is historically possible, scientifically realistic, and socially conceivable in our time: the establishment of a balanced picture of nature and man acceptable to all peoples and covering all the knowledge accessible today either to the religious intuition or to the scientific intellect. This could achieve much if it provided every man with a valid symbol of the constructive character of his own life impulses and the productivity of his own mind.

We are here in the world of hard facts, true potentialities, and possible achievements. Nothing in human knowledge of nature or in human history excludes the possibility that a formative principle may provide the clue to (1) the needed theory of the fundamental particles; (2) the understanding of biological organization; and (3) the working of the human brain. It is unusual that the future development of science should be anticipated in this manner. But why not, if the need is so great? The way can be prepared by pointing to a possibility.

A more important question is why this possibility has not been recognized sooner, if it is valid.

The reason lies in a bias in the scientific intellect which until now has been necessary. This may be summed up in a triple emphasis on *analysis, permanence,* and *abstraction.* From the origins of science the most fertile path of advance has hitherto lain in the search for permanent parts attainable by a process of abstraction from immediate observations. There is therefore no cause for surprise that a principle which involves a partial reversal or compensation of these three tendencies has lain well hidden. For the conception of a formative process puts the accent not on abstract permanent parts, but on the immediacy of changing forms.

Science itself could benefit from a fuller recognition of the unconscious preferences which have guided its historical development and still persist today. Einstein has put on record his view that in some the devotion to exact science expresses a flight from personal frustration into a world of numerical harmony, and that scientific theory must advance by continually increasing abstraction. The motive expressed in the first suggestion may explain what may be the error of the second. For we must hope that it was an error and that a long period of increasing abstraction will now be balanced by a return toward immediacy. Even if all intellectual endeavors constitute a temporary desertion of personal situations, that need not imply a flight into abstraction so desperate as to involve the neglect of transformation and history and the acceptance of the Pythagorean illusion of a static harmony. Certainly the anticipated development in science requires as its instruments a kind of scientist who accepts his personal situation in a world of transformation and expresses this recognition of change in his scientific thought.

It seems that the chief need is greater immediacy in thought.

Immediacy is difficult to achieve. Once the sails of the mind have taken the winds of abstraction, we are lost in a Pythago-

rean dream and are condemned to ultimate frustration. Our
vitality can only be adequately expressed, and the deeper
natural order discovered, by a return to immediacy.

This was not always so. There was a long period in the
history of the human mind in which further abstractions were
necessary and effective. But that voyage away from the immediate facts may be near its end. We must now return to
ourselves and our experience. Social health, personal balance,
and intellectual order can only be recovered by bringing
thought and experience into closer relation to each other and
to the vitality from which both spring.

The recovery of immediacy in rational thought implies its
restoration from a condition of abstraction and isolation to its
true place as a phase in the transformation of feeling into
action. When thought is completely abstract its relation to
the organic situation is lost. It has to be brought back into
place so that we can visualize our thoughts, feel them in their
relation to our own impulses, and live them out in action.*

The German language has a word, *anschaulich,* weakly
translated as "visualizable." Thought must not merely be
anschaulich, but become an integral part of life, recognized as
springing from vitality and expressed in positive affirmations
and actions.

The contemporary sense of loss of coherence, of the disorganization of human existence, is not merely the con-
comitant of a time of rapid change. It is also the expression
of a disorganization of our rational awareness.

This failure of coherence is three-fold:

Thought itself is internally disorganized; there are no

---

* "With Zeus alone, thought, word, and deed are one." Aeschylus,
The Suppliants.

elegant ordering principles to guide thought along clear channels through the diversity of experience.

Thought has no direct relation to the immediate experience of being alive, to the one-way flow of process, the rhythms of life, the transformations within and around us.

And finally thought does not flow unchecked into action; thought is deeply and essentially frustrated.

Thus thought, which should be elegant in its own structure, integral with experience, and continually flowering in action, is none of these. This is the wrongheadedness of our age, and the remedy may lie in an adjustment restoring the immediacy of thought, as the expression of vitality, the mirror of experience, and the fountain of action. To achieve this we must place the accent on form.

XII.

The Magic of Form

Beauty is in the forming power

A HEALTHY creature cannot repudiate its own vitality. The activities of a harmonious being must be an affirmation of its life impulse. Only a dissociated man can feel that life is not worth living.

Our dilemma is that we know this and yet many of us cannot make an appropriate affirmation. We desire personal and social harmony, we want to believe in ourselves and our fellows, but we distrust human nature and remember the German gas chambers and the persistence of war. How can so scarred a generation rejoice in living?

An earlier kind of man was able to separate the conflicts of this world from the harmony of another, and to live contentedly here dreaming of the other. Pythagoras, Plato, St. Paul, and a great part of the Western world found this way of life satisfactory.

But a change is taking place in the human state of awareness. We are now too vividly conscious of our own thought processes to enjoy a naïve confidence in a harmony elsewhere which may be only a psychological compensation for otherwise unbearable frustrations. We require a new kind of honesty so that we can see everything as it is and yet make the best of it. The question is: What kind of positive affirmation regarding human life is compatible with honesty?

If one asks which Western personalities, say since 1600, expressed in their life and thought a vigorous affirmation of human existence, the answer is bleak. There are not many of the first rank who merit consideration. Of these Spinoza, Shaftesbury, Goethe, Emerson, and Albert Schweitzer come to mind. All these in some manner affirmed their vision of god-nature-man, of an underlying unity which is good. But the majority of Western thinkers did not feel or think like that.

To an unprejudiced mind contemplating the Western world afresh this must appear as a significant fact. Is Western man a kind of being who rejects his own vitality? If so, have these few men been adequately valued for their unusual personal condition and the consequent originality of their thought? One would imagine that a morbid community might pay attention to prophets of health. But who reads their writings for this? Are they praised in the schools and colleges for their rare acceptance of life? Or do we really think it foolish to accept and affirm human life?

As I have written about Goethe elsewhere, and as Schweitzer is too near us, Spinoza too ascetic for my liking, and Emerson provides no suitable quotation, I shall take a passage from Shaftesbury to illustrate a comprehensive affirmation. It is from *The Moralists; A Philosophical Rhapsody,* by Anthony Ashley Cooper, Third Earl of Shaftesbury, 1671–1713, philosopher, moralist, and aesthetician:

> Here then, said he, is all I wou'd have explain'd to you before: "That the Beautiful, the Fair, the Comely, were

never in the Matter, but in the Art and Design; never in Body it-self, but in the Form or forming Power." Does not the beautiful Form confess this, and speak the Beauty of the Design, whene'er it strikes you? What is it but the Design which strikes? What is it you admire but Mind, or the Effect of Mind? 'Tis Mind alone which forms. All which is void of Mind is horrid: and Matter formless is Deformity it-self.

Of all Forms then, said I, Those (according to your Scheme) are the most amiable, and in the first Order of Beauty, which have a power of making other Forms themselves: From whence methinks they may be stil'd the forming Forms. So far I can easily concur with you, and gladly give the advantage to the human Form, above those other Beautys of Man's Formation. The Palaces, Equipages and Estates shall never in my account be brought in competition with the original living Forms of Flesh and Blood. And for the other, the dead Forms of Nature, the Metals and Stones, however precious and dazzling; I am resolv'd to resist their Splendour, and make abject Things of 'em, even in their highest Pride, when they pretend to set off human Beauty, and are officiously brought in aid of the Fair.

Do you not see then, reply'd Theocles, that you have establish'd Three Degrees or Orders of Beauty?

As how?

Why first, the dead Forms, as you properly have call'd 'em, which bear a Fashion, and are form'd, whether by Man, or Nature; but have no forming Power, no Action, or Intelligence. Right. Next, and as the second kind, the Forms which form; that is, which have Intelligence, Action, and Operation. Right still.

Here therefore is double Beauty. For here is both the Form (the Effect of Mind) and Mind itself: The first kind low and despicable in respect of this other; from whence the dead Form receives its Lustre and Force of Beauty, For what is a mere Body, tho a human one, and ever so exactly fashioned, if inward Form be wanting, and the Mind be monstrous or imperfect, as in an Idiot, or Savage? This too I can apprehend, said I; but where is the third Order?

Have patience, reply'd he, and see first whether you have discover'd the whole Force of this second Beauty. How else shou'd you understand the Force of Love, or have the Power of Enjoyment? Tell me, I beseech you, when first you nam'd these the Forming Forms, did you think of no other Productions of theirs besides the dead Kinds, such as the Palaces, the Coins, the Brazen or the Marble Figures of Men? Or did you think of something nearer Life?

I cou'd easily, said I, have added, that these Forms of ours had a virtue of producing other living Forms, like themselves. But this Virtue of theirs, I thought, was from another Form above them, and cou'd not properly be call'd their Virtue or Art; if in reality there was a superiour Art, or something Artist-like, which guided their Hand, and made Tools of them in this specious Work.

Happily thought, said he! You have prevented a Censure which I hardly imagin'd you cou'd escape. And here you have unawares discover'd that third Order of Beauty, which forms not only such as we call mere Forms, but even the Forms which form. For we our-selves are notable Architects in Matter, and can shew lifeless Bodys brought into Form, and fashion'd by our own hands: but that which fashions even Minds themselves, contains in it-self all the Beautys fashion'd by those Minds; and is consequently the Principle, Source, and Fountain of all Beauty.

It seems so.

Therefore whatever Beauty appears in our second Order of Forms, or whatever is deriv'd or produc'd from thence, all this is eminently, principally, and originally in this last Order of Supreme and Sovereign Beauty.

Here is a splendid Yea-saying expressed in the language of form. Shaftesbury was deeply influenced by classical literature, and yet he was able to anticipate the emphasis on process and development which after his death was to dominate European thought: beauty not merely in the form, but in the forming power.

This passage displays a fusion of Platonism and process, of classical idealism and joy in fertility, that is rare and we may hope prophetic. For some of Plato's errors have here been discarded. Full place is given to the beauty of the human figure, to the living forms of flesh and blood with their force of love and power of enjoyment, to the forms which have the power of making other forms, and to the supreme principle which fashions the forming forms. The Platonic dualism has given place to a real hierarchy of living beauty, alive in this world. We rejoice in the present process, not merely in past or future.

This appears to me very like the face of the new god. Behind Libido and Life, more comprehensive than either, we recognize Shaftesbury's universal Forming Power.

For Shaftesbury there are three orders of beauty in ascending value: first, the dead forms; then the forming forms, with intelligence, action, and operation, that can produce other forms; and finally the sovereign principle which is the creator of all forms, the source and fountain of all beauty.

Could anything be happier than that Shaftesbury, who found the key to morality in aesthetic criteria, should have so superbly anticipated what must surely be the central conception of any future all-in wisdom, whether we call it religious or scientific? The dead forms call for no comment, though we understand them better than he did. The forming forms are all the organic structures and systems, from the ultimate self-reproducing protein units to the human person, that can make new forms both by sexual reproduction and by imaginative creation with his hands or brain. And the sovereign principle which creates all forms, for Shaftesbury the Divine mind not merely passively pervading all nature but actively shaping it, is for us, or will be for our children, the universal formative tendency which is the source of everything that possesses form. I believe that Shaftesbury's philosophical rhapsody was, and will be recognized to have been prophetic.

One may regard with suspicion the asceticism of the Pythagorean brotherhood, but if Shaftesbury had established a community sworn to honor his supreme source of beauty, the forming forms, and the dead forms, many would have found in it their spiritual home. Indeed Shaftesbury did create an invisible brotherhood, for his seeds of thought germinated in many minds both in England and on the continent of Europe during the eighteenth and nineteenth centuries. One may go further and dream, perhaps not altogether foolishly, of an invisible community already in being, living not by any collective doctrine, but by their personal sense of the forming power in inanimate nature, in organisms, and in themselves. One can imagine a marriage of pagan fertility and Pythagorean harmony in a reverence for this forming power. The long Platonic-Christian austerity and denial has accomplished its purpose and can now surrender to living beauty. The cold ecstasy of idealism is exhausted, a natural rhythm of fulfillment can take its place. For today we can recover the knowledge that acceptance of our own vitality need not bring with it relapse into a damaging sensuality, but can promote the natural enjoyment of all powers, natural, organic, and creative.

And yet Shaftesbury's rhapsody is unreal. It is like a restful dream of another and better world, a Mozart harmony borrowed from the past, whereas what we experience in this century is the world of struggle portrayed by Beethoven and the moderns. Donne's cry: " 'Tis all in pieces" is ours, rather than Shaftesbury's praise of a state of innocent well-being.

Shaftesbury's optimistic rhapsody was written exactly a hundred years after Donne's despairing "Anatomie." Donne looked back and saw in his own time only the corruption of a past world which he admired. Could he have looked forward and foreseen Shaftesbury's mood? No. And can we foresee an experience of happy living in the decades to come, anticipate in some degree the quality of the affirmation which our children and grandchildren must make if civilization is to survive? Perhaps we can, to the extent that our temperament and imagination permit.

The problem is how to make Shaftesbury's affirmation though experiencing Donne's disruption. Is it merely flattery to our age to recognize that we are now faced with the necessity of a new kind of honesty? We have to make a whole-natured affirmation of human life, while recognizing the inescapability of frustration not only for ourselves but for others.

The affirmation must be of a principle pervading all nature and known to each of us in our own nature, which enables us to face distress and destruction. It should be so general and so deep that we are not shaken by any particular disaster. The affirmation of a universal formative process, if that is the shape it takes, must be so well grounded in both subjective experience and scientific knowledge, that no destruction of individual forms can disturb it. The supreme beauty has to be recognized in the general forming power, rather than in any particular form. If that recognition could be achieved and sustained, nothing would shake us.

Here our perspective of the coming years can become universal rather than merely Western. For at this point the Western tradition recognizes the validity of an ancient doctrine of the East: the universal principle has to be valued above any particular expression, if serenity is to be achieved.

Many of the finest passages in Taoist writing express a lyrical acceptance of nature and life, as realized in the condition of Tao. It is broadly true that Eastern thought, like much German philosophy, emphasized the sense of a process of change pervading everything and tended relatively to neglect the specific forms of individual things. This quasi-mystical vision of a universal process was challenged by the developing Western intellect, which laid stress on the recognition of particular forms and their differences. But the Western preoccupation with individual forms and differences has reached a critical point where a more balanced view is necessary if the tradition is to survive.

Neither the Eastern emphasis on an undifferentiated unity of process nor the Western stress on specific differences can carry the human tradition through this century. The time has come for a new elegance: a unity of process seen in all particular forms and reconciling their differences. A fresh stress must be laid on universal principles in order to restore a proper equilibrium.

But once again there is no balance in the dialectic. East

and West are not on an equal footing. The West has made the modern world and the West must redeem it. Until Western science recognizes the formative process no fundamental advance is possible.

The supreme beauty is not in the form, but in the forming power. The deepest aesthetic and scientific principle lies in a tendency toward simplicity, order, elegance, form.

Take one example: the life of a man. A fertilized ovum, by a continuous but structured process, becomes an adult, fulfills itself, and is dispersed.

How little we know or understand about this process! I may think I know a young man called John. But do I? For John is one cell that has unfolded into a human being, and I do not know how that happened. John is a hereditary potentiality realized in a particular environment, in an unknown manner.

I may like John and feel sympathy with his desires and aspirations, but do I understand him? Do I see the latent potentialities which have never been realized, the half-developed faculties which have been thwarted, the characteristic situations which haunt his life, the secret distresses which John experiences but on which he himself does not reflect since they lie too deep and are too sensitive to be thought about?

No one has ever fully understood one living human being with a sympathetic insight, at once intuitive and rational, into the manner in which physical, organic, mental, and social factors have interacted to form the person. Here is the forming power at its highest, not yet understood.

If one could achieve this insight one would experience the full beauty of the forming power, the majesty and necessity of a hierarchy of formative processes, the mingled story of satisfying fulfillment and painful frustration. And one would be able to recognize at every moment in every human life the still sensitive growing point, the potential source of new development.

The awkward feature, which one can do nothing about and must accept, is that in any particular human being the growing point may *for practical purposes* be dead.* There is no general law of universal validity that guarantees in any particular case the possibility of an advance toward harmony, even given the optimum conditions. There are inertias, twists, treacheries, poisons, and disintegrations of personality which have deep roots and cannot be overcome by anything within human power today. The beauty lies in the general forming power, in the universal tendency, rather than in any particular form.

There is no ground for surprise in the existence of "evil." We may be ready to fight Hitler, but the fact of his existence and of his following need not startle us. The chances of heredity, the risks of human life, and the absence in particular historical periods of an appropriate way of thinking, render it inevitable that human character must suffer from a wide range of distortions. The fact that we experienced astonishment at the emergence of the situation Hitler-ruling-Germany shows that we had illusions.

Mass hysteria, cancer, and psychosis offer no cause for astonishment, for it is easy to understand so subtle a thing as biological organization breaking down, particularly in the sensitive human brain and in unstable human communities.

---

* The real death is the gradual cessation of growth, the fading out of the forming power, not the end of the physiological cycles.

But there is one fact which is astonishing, that ought to serve as compensation for all the negative aspects of existence.

This is the basic harmony in the human organism. For example, the development which takes place from the single parent cell to the newly born infant displays a single harmonious unfolding of interrelated processes and structures so extraordinary as to defeat all comment. That mischances sometimes occur is natural, but what could not be expected on the basis of any known scientific theory or philosophy is the stupendous degree of harmony or organization, the unity expressed in so much diversity, the co-ordination of so many organic principles and functions in the embryo and infant that is on his or her way to become a unique human person.

Lenin held that the cry of a single child in distress condemned the world. But only if a sentimental and irrelevant conception of justice to the individual is the criterion. The world is not made that way, and we shall hate life and be cruel to our fellows if we rage at finding ourselves in a mode of existence where there is no such justice. We were never asked if we wanted to be born, but that does not condemn existence.

On the contrary the successful delivery of a single healthy child redeems nature. For it shows that deep in the structure of this universe in which we find ourselves there is a powerful principle which works toward harmony, which molds and sustains and develops so intricate, subtle, and specific a thing as a human embryo, and allows it to become a man or a woman.

The healthy baby enters on his life. He does not know and will at first misunderstand the nature of his fate. He belongs to a species that is not in general capable of retaining harmony in adult life without an elegant way of thinking which allows the individual to see the whole without neglecting the details. This he needs for biological adaptation; he has not got it; his life will therefore be more painful than it would be with it.

One might say that this was the supreme injustice. For if such a way of thinking were available no disease would be incurable and no pain unbearable. But there is nothing that can be done about it, except to try to create that way of thinking. The idea of a universal formative process is a contribution toward that elegant view of things. It is a germ that still has to unfold.

Shaftesbury liked to use symbolic illustrations to give immediacy to his argument and to put them on the frontispiece of his books. I have not found the visual image which can suggest the forming power of nature in all its guises.

Perhaps there cannot exist any one adequate symbol of a power that is so rich. If you suffered a long illness and had to lie in the dark, how would you occupy your imagination?

One might try to summon before the mind's eye those images that were richest in significance in relation both to human experience and to the order of nature.

One might recall the nude human figure, at all ages; the human face as an instrument for expressing and conveying internal conditions; the human skeleton; human sperm under the microscope; Leonardo's drawings on "Generation"; Michelangelo's unfinished sculptures; a serene face in old age; the Mexican pyramids seen from the air; neolithic cave drawings almost neglecting man himself. . . .

And now, at this moment, if I demand from memory further images, what arrive? The face of a friend; a favorite picture; a lamb being born and learning to walk; the surface of the moon; the strange forms of the nebulae; and so it goes on.

Different images will come to you.

These fixed images do not spring into existence without good cause, they are each the record of a process which formed them. However significant they are, perhaps you will agree with Shaftesbury that the supreme beauty is in the forming power. For without that there would be nothing, no order, no life, no humanity.

Follow Up

The following notes may be helpful to anyone who wishes to pursue further the argument of this book.

An extensive bibliography and chronological survey on Form was published in *Aspects of Form: A Symposium on Form in Nature and Art*. Edited by the author (New York: Pellegrini & Cudahy, 1952). There has since appeared the valuable work by Hermann Weyl: *Symmetry* (Princeton: Princeton University Press, 1952).

On the general historical background see the author's *Next Development in Man* (New York: Holt, 1948), and on the physical and biological aspects his *Unitary Principle in Physics and Biology* (New York: Holt, 1949).

Susanne Langer's *Philosophy in a New Key* (Cambridge, Mass.: Harvard University Press, 1942) is warmly recommended on symbolic forms in all branches of culture.

The following more specialist works are relevant to the argument and have been used:

H. C. Urey, *The Planets, Their Origin and Development* (New Haven: Yale University Press, 1952); R. B. Baldwin, *The Face of the Moon* (Chicago: Chicago University Press, 1949); F. E. Zeuner, *Dating the Past* (London: Methuen, 1946); R. J. Pumphrey, *The Origin of Language* (Liverpool, 1951); and N. Tinbergen, *The Study of Instinct* (New York: Oxford University Press, 1952).

A popular treatment of the micro universe is given in *Mr. Tompkins Learns the Facts of Life*, by George Gamow (Cambridge University Press, 1953).

Those who desire a more rigorous formulation of the main ideas may find the following definitions helpful:

Formative process: a process represented by the decrease or vanishing of a positive parameter measuring the deformation of a complex system from some stable extremal state. This can be regarded as a more precise definition of "a process leading from disorder toward order."

Formative theory: A theory which represents change not as the *relative motion* of pairs of objects, or as the *increase of entropy* in statistical systems, but as the *formation and deformation* of complex arrangements of *n* particles.

Form: The terminus of any formative process, requiring fewer independent parameters to represent it than earlier states, that is, a well-ordered arrangement.

Structure: Any form, or formative process, considered in its full detail.

Fundamental physical structure. The changing three-dimensional spatial relations of a system of *n* points representing permanent particles. Transient particles would represent states, or transitions between states, of systems of these permanent particles.

On problems in Eskimo point arrangements see *Am. Math. Monthly,* vol. 56, p. 606, 1952.